COUNCIL *on*
FOREIGN
RELATIONS

Maurice R. Greenberg
Center for Geoeconomic Studies

Impact of Climate Risk on the Energy System

Examining the Financial, Security, and Technology Dimensions

Amy Myers Jaffe, Joshua Busby, Jim Blackburn, Christina Copeland, Sara Law, Joan M. Ogden, and Paul A. Griffin

CONTENTS

INTRODUCTION

Amy Myers Jaffe and Joshua Busby

The effects of climate change pose risks not only to the earth's natural ecosystems but also to the security and livelihood of the people of the United States and around the world. In its 2018 special report, the Intergovernmental Panel on Climate Change (IPCC) suggests that the speed and scale of the consequences of global warming have intensified faster than projected. The IPCC concludes that in some parts of the globe, at certain times of year, temperatures have already risen 1.5 degrees Celsius above preindustrial levels, increasing the probability of severe weather and other conditions expected from a warming planet. Still, specific challenges vary by location and intensity, even within national geographies such as the United States. Thus discussion that focuses on global average measurements and consequences will miss the mark.

The recent U.S. congressionally mandated Fourth National Climate Assessment, released in 2018, concludes that climate change will increasingly threaten the U.S. energy supply via more frequent and longer-lasting power outages that will broadly affect critical energy infrastructure. The assessment is prepared by the U.S. Global Change Research Program, which includes contributions by the National Oceanic and Atmospheric Administration (NOAA), U.S. Department of Defense (DOD), U.S. Department of Energy (DOE), and several other U.S. government agencies. Some regions, such as the West and Southwest, have witnessed compounding effects such as droughts combined with high temperatures and wildfires. Other regions, such as the U.S. coast of the Gulf of Mexico and the Eastern Seaboard, are experiencing severe storms combined with high temperatures and flooding. Other parts of the country, including the Midwest, have seen extreme rainfall.

Consequences of climate change affect virtually every aspect of the U.S. energy system. As climatic effects such as rising seas and extreme weather continue to make themselves painfully obvious across many geographies, U.S. energy infrastructure is increasingly at risk. The United States is ill prepared for this national security challenge. Climatic disruptions to domestic energy supply could be large, entailing huge economic losses and potentially requiring sizable domestic military mobilizations when blackouts and water and fuel disruptions create health and safety emergencies among civilian populations. Yet public debate about emergency preparedness is virtually nonexistent.

To explore the challenges of climate risk to the U.S. energy system, the Council on Foreign Relations organized a two-day workshop on climate risk to U.S. energy systems and national security in New York on March 18 and 19, 2019. The gathering of fifty participants included current and former state and federal government officials and regulators, entrepreneurs, scientists, investors, financial- and corporate-sector leaders, insurers, members of credit agencies and nongovernmental organizations, and energy policy experts. During their deliberations, workshop participants explored how climate-related risks to U.S. energy infrastructure, financial markets, and national security could be measured, managed, and mitigated. To guide the discussion, participants produced five essays, reproduced here, on topics related to the financial, technological, and security dimensions of climate risk to the energy system.

The workshop program began with discussion of the physical climate risks that will increasingly be felt in different parts of the United States and present a growing problem for U.S. energy security, the U.S. economy, and U.S. national defense. Energy facilities, military bases, and communities on the U.S. coast of the Gulf of Mexico are particularly exposed to sea-level rise. This is partly due to changes in expected sea currents combined with local land subsidence (the sinking of ground through natural erosion and the removal of underground materials such as water, oil, and gas during commercial oil development). Scientists estimate that the region could experience up to four feet of additional sea-level rise by 2100. This makes Gulf Coast refining, which constitutes 44 percent of total U.S. refining capacity, highly vulnerable to flooding events and dangerous ocean surges during severe storms and hurricanes. Refineries on the Gulf Coast serve the entire country through a connected network of nearby pipelines that are critical to transporting gasoline, jet fuel, and heating oil across the entire United States. Several of the nation's largest ports, which host the majority of terminals used for exporting U.S. oil and natural gas to global markets, are also located on the Gulf Coast.

Financial markets have been slowly incorporating information about climate change into valuation of some products, but the fate of energy company stocks, bonds, and oil and gas commodity derivatives is uncertain. The Bank of Canada, Bank of England, European Central Bank, and Norwegian Government Pension Fund Global, among others, have expressed concerns that more systemic risks could loom if unexpected changes in valuations come about quickly. Markets could experience a cascading effect if physical damage to corporate assets and facilities, legal liabilities, or regulatory risks emerge suddenly.

Even in the aftermath of the bankruptcy of California utility PG&E Corporation, whose faulty equipment could be found responsible for several deadly wildfires last year, rating agencies have not sufficiently downgraded credit to the U.S. utility sector. Bond rating agencies have begun to downgrade cities based on frequent climate events, but they apply less rigorous climate ratings procedures for credit risk to electric utilities and other energy companies. Large pension funds and some other sizable institutional investors have tried to engage energy companies with demands to increase transparency on how the companies plan to incorporate climate risk into investment decision-making. But throughout the United States, power-generation companies still have unfettered access to cheap financing, regardless of known climate-related risks to physical facilities. Securities valuations similarly do not appear to reflect the full risk that the commercial net worth of carbon-intensive assets—such as coal mines, proven oil and gas reserves, and related processing facilities—could depreciate unexpectedly or become stranded assets (that is, become obsolete because of a transition to cleaner energy sources). Investors and credit analysts still assume that companies will be able to recover their returns on capital expenditure in the ten-to-twenty-year time horizon typically under analysis. This sanguine view ignores the possibility that sudden changes in valuations could take place. Already, some energy producers and electric utilities have seen their cash flows hampered after facilities were damaged in storms or wildfires. Similar cases are likely in the future. A court ruling or new environmental legislation can also change the cash flow expectations for businesses. Over the last seven years, the market capitalization of U.S. coal firms has fallen from $62 billion to under $10 billion. Cloud Peak Energy, the third-largest coal company in the United States, filed for bankruptcy in May.

The U.S. Securities and Exchange Commission (SEC) does not currently ensure that disclosures of material risks related to climate change are accurate or sufficiently detailed. Company filings to the SEC that mention carbon-related transition risks typically use boilerplate, generalized language that is not company specific. The SEC received an active slate of shareholder proposals for more disclosures on climate risk this year but dragged its feet on initiating any guidelines beyond those provided in 2010. The agency lacks the funding and resources to generate guidance rules internally and often sides with companies that claim they have no material risks to disclose. The agency is said to be seeking a market-driven way to guide any future formal rulemaking, but so far its response has been slow as it remains bogged down in debate on the best way forward.

The current dire situation of PG&E offers a window into kinds of risks that can emerge suddenly and with little warning from rating agencies or financial analysts. California's courts have ruled that under the principle of inverse condemnation, PG&E is strictly liable for any wildfire damages linked to their equipment because the company destroyed life and property while performing its public function. PG&E, which still serves sixteen million customers in Northern California, now has little access to capital markets, which constrains its ability to invest in future infrastructure or even to inspect, repair, and upgrade existing facilities—which could cost from $75 billion to $150 billion. Utilities in California used to be able to access both direct insurers and reinsurers to cover their potential liabilities. Now premiums are rising, and state officials and business leaders are concerned that private insurance markets could fail. The State of California has set up a commission to investigate how to insure against these climate risks. However, the funding sources for catastrophic risks such as wildfires are unclear.

Public-sector options are needed, both to mitigate physical risks to U.S. energy installations and to bolster private insurance markets that can ameliorate the financial consequences of the more extreme risks. But so far, lessons from the private U.S. real estate market are not encouraging. Canada is ahead of the United States in thinking about how to ensure properly functioning insurance markets, considering cases where sectors or localities could become uninsurable. Widespread residential flooding in Canada presented policymakers and insurers with a massive problem that threatened the proper functioning of an insurable housing market, which led policymakers to intervene to try to incentivize the reduction of climate-related risks to housing developments via the federally funded National Disaster Mitigation

Program (NDMP). The NDMP focuses on infrastructure investments that address the costs and risks of recurring residential flooding with an eye to facilitating the proper functioning of private residential insurance markets. Some form of climate change prediction markets could provide aggregated information for investors and add transparency to how market participants price the probability of scientific projections, such as future sea-level rise or frequency of heat waves.

In "Climate Change, Storm Surge, and the Oil and Gas Industry," Jim Blackburn and Amy Myers Jaffe argue that state and federal authorities should work together to update building standards for seawalls, levees, and storage tanks to reflect scientific projections for future risks. Regulators in states along the Gulf of Mexico should require refineries and the petrochemical industry to compile and maintain a current inventory of hazardous chemical volumes that are stored on-site or near their facilities. The essay highlights the clear risk of serious inundation to refining capacity along the Houston Ship Channel, as well as the risk of damage to as much as 50 percent of the 4,400 petroleum and hazardous waste storage tanks also found there. Texas has not yet tackled how damages from the many spills that took place during Hurricane Harvey will be addressed or who will pay the costs. Similar catastrophes could become more frequent in the future and the authors urge Gulf Coast state authorities to consider how to set up permanent disaster-response funding that can go beyond individual claims for compensation for toxic releases to other uses, such as funding restoration of soil, waterways, and other impaired ecosystems. Precedents such as the Deepwater Horizon Oil Spill Trust and the 1980s Superfund program can serve as starting points for study of an appropriate funding structure. The essay also covers vulnerabilities in California, where refineries could become exposed to flooding as seas rise. It suggests that California consider how to build redundancy into its fuel transportation system, including pre-positioned fuel inventories for first responders.

The link between water availability and energy production creates another layer of risk to U.S. energy security. Almost all forms of energy production, including electricity generation, require a stable supply of good-quality water. Droughts can disrupt the electricity system by decreasing availability of hydroelectric power or curtailing thermal and nuclear electric-power generation, which depend on water for cooling operations. Climate change is precipitating significant changes to water quantity and quality in the United States. Rising temperatures are intensifying droughts and reducing snowpack, threatening to alter water patterns in places where the U.S. energy industry taps substantial volumes of fresh water in its daily operations. In their essay, "Water-Related Risks and Impacts on the U.S. Energy System," Christina Copeland and Sara Law argue that unmanaged water risks could have costly outcomes for the energy industry, in both financial and operational terms. Higher operating costs, costlier financing and insurance, constraints on growth, and stranded assets are all consequences for energy producers that could stem from climate-related water insecurity. Water-related issues have already emerged as a risk to energy production in certain parts of the United States. Given such circumstances and future risks, companies and local governments should undertake water-related risk assessments and consider technologies and techniques that can minimize water-risk exposure. The authors point out that energy companies vary greatly in how they manage water risk. They estimate that in 2017 alone, twenty large representative energy companies experienced detrimental water-related disruptions to their operations, totaling $1.8 billion in revenue losses, due to water scarcity.

Because other critical infrastructure, such as U.S. refining and water transportation and supply, relies on electricity services, electricity outages have the potential to take fuel production and distribution sites and even retail gasoline stations offline. In her essay, "Climate Change Impacts on Critical U.S. Energy Infrastructure," Joan M. Ogden discusses the risk of these cascading effects, which were chronicled in detail in the Fourth National Climate Assessment. In parallel with climate change, the U.S. energy system is in the midst of a major transition driven by technological advancement, economics, and policy. Ogden recommends designing the future system to be inherently resilient to the worsening climate changes expected later this century, even with strong measures to cut emissions. For example, some localities have found that micro-grids can be restored more easily than large centralized thermal plants after an extreme event such as a hurricane. States that have experienced climatic event–related disruptions to the electricity needed to pump gasoline into distribution trucks and pipelines now require the local suppliers of that fuel—for example, motor-fuel terminals and wholesalers—to have portable backup generators available. Ogden argues that options will take many forms, depending on regional conditions and resources. For example, New York is exploring how to use buildings as virtual power plants that aggregate electric power from rooftop solar, electric cars, and building-level battery storage and deploy the excess capacity to the wider grid. Such systems are more likely to bounce back quickly from severe weather events. Ultimately, Ogden concludes, innovative public and private financing mechanisms are needed to promote policy options.

Financing new energy infrastructure and necessary climate change adaptation measures has proven difficult in the United States. In the essay, "U.S. Climate Risk and Financial Markets," Paul A. Griffin and Amy Myers Jaffe argue that routine underpricing of climate risk affects firm-level financial valuations. Firms are likely subject to higher and more uncertain future net cash outflows than is currently being taken into account because of the costs of adaptation, innovation, and mitigation, not only in public-equity stock markets but also in analyses by rating agencies, insurers, and other financial institutions. Uncertainty about the nature and timing of future regulations, the amount and timing of loss of cash flow related to physical climate risks, and energy-fuel transitions is creating constraints that hinder the efficient assessment of climate risk. The essay warns that if investor inattention leads to climate risk underpricing, it could increase the propensity for herding (when investors make decisions based on the actions of other investors, and not on accurate information) that could destabilize markets in the future. The essay suggests that though disclosures regarding weather-related risks are becoming more common among U.S. publicly traded firms, corporate disclosures lack the level of detail investors need to assess firms' strategies and investment plans to meet states' or anticipated national or global climate change adaptation and mitigation goals. The essay asserts that the SEC should revisit its almost-ten-year-old guidance statement on climate change.

Climate risk could manifest not only in physical damages but also in financial market failures or cascading energy shortages, which highlights the complexities of the threat of climate change to U.S. national security. The DOD has carried out a number of studies to assess the vulnerability of its military bases to climate harms and how climate change could affect military missions, operations, and training. But the threat that climate change poses to U.S. national security goes beyond force preparedness and deployment. As Joshua Busby argues in his essay, "A Clear and Present Danger: Climate Risks, the Energy System, and U.S. National Security," the severity and speed of potential harms present security threats in their own right. Extreme weather events increasingly require military mobilization to prevent loss of life, deliver relief, reestablish order, and restore energy and other critical services. Moreover, the uneven distribution of climate disasters could also damage confidence in the effectiveness or fairness of U.S. political leaders or worsen political divisions if affected constituencies' confidence in the system is undermined.

The DOD should also consider the risks that climate change poses to the military's energy supply. The U.S. military is the largest customer of the U.S. electric grid, which is operated mostly by the private sector. Vulnerabilities extend beyond physical military facilities to the surrounding communities that house families of service members and support external vendors that are also important to military preparedness. Many other critical infrastructure systems rely on electricity services: electricity outages that result from high winds, extreme temperatures, storms, or wildfires can take vital infrastructure, such as hospitals, transport systems, and data centers, offline. The consequences for civilian populations would be severe and would likely require military mobilization to help them. Energy resilience should be a prime element of the DOD's core readiness preparation.

In light of the increasing risks to U.S. energy production and supply from changing climate systems, the United States should better prepare for climatic events, reduce risks, and mitigate consequences through using new energy technologies, financial tools, and changes to disclosure regulations. The first step in this process is to improve the public's understanding of the detailed regional scientific assessments of future patterns of heat waves, sea-level rise, droughts, and wildfire risk under different global warming scenarios. Lawmakers should ensure that the latest scientific assessments of regional challenges to critical infrastructure are shared with the U.S. Army Corps of Engineers, the DOD, and the Federal Emergency Management Agency (FEMA) as a basis for planning capital expenditures for adaptation, mitigation, and evacuation.

RECOMMENDATIONS

Climate risk to U.S. energy infrastructure represents a major threat to the U.S. economy and national security. Policymakers should give more careful attention to policy options to identify, anticipate, and mitigate this risk. The role of federal, state, and local governments is critical, but the private sector also needs to offer innovative financial and technological options that can distribute the costs of preventive action efficiently and equitably across society. The following policy recommendations address the gaps in knowledge that thwart more effective responses.

- *Congress should require the Department of Homeland Security and FEMA to update risk-assessment mapping by geographies, infrastructure type, and specific climate hazard such as drought, heat wave, flooding, or severe storm.* Those agencies could then identify future climate-related vulnerabilities to the U.S. energy system, in general, and to energy supplies to U.S. military bases and operations, in particular. The assessments should include corollary issues such as energy supply to nearby support infrastructure, including data centers, hospitals, and food production sites. Congress and state governments should use the data to assess and initiate infrastructure projects and emergency readiness procedures that address damage from climate change.

- *The U.S. Army Corps of Engineers and FEMA should commission and conduct local and regional scientific studies to define future patterns of heat waves, sea-level rise, droughts, and wildfire risk to critical infrastructure, including energy and water resources and facilities.* These organizations should use the studies as the basis for planning capital expenditures for adaptation and evacuation. State governments and the National Science Foundation should increase funding to regional university centers to complete these assessments, which should include a range of probable global-warming scenarios.

- *States and the federal government should work together to update building standards for seawalls, levees, and storage tanks to reflect accurate scientific projections for future risks and address the specific vulnerabilities of Gulf of Mexico refineries, terminals, and energy distribution infrastructure.* Regulators in U.S. states along the Gulf of Mexico should require refineries and the petrochemical industry to compile and maintain an inventory of hazardous chemical volumes that are stored on-site or near their facilities.

- *Credit agencies should consider risk from climatic events when calculating credit ratings to improve the analysis of risks to profitability and performance.* They should factor in risks to corporate inventories of hazardous chemicals that are stored on-site or near energy production; corporate histories of environmental performance; and safety records, including spill records and any histories of fires and explosions related to faulty equipment.

- *States and the federal government should standardize and tighten water-usage reporting and risk-disclosure requirements for corporations and utilities.* Companies and communities cannot address water scarcity risks that could affect energy production if they do not have access to accurate data about how much water industry and local consumers use. Reported data could be used to promote sustainable water practices in regions that could face water shortfalls. An assessment of local requirements could facilitate cooperation in water stewardship for the region by energy companies, communities, and regulators. Such cooperation can be critical to prevent water shortages from triggering energy and electricity outages or supply constraints.

- *The SEC should establish permanent disclosure standards for climate risks to the operations of publicly traded energy companies and utilities.* To start, the SEC should fully participate in fact-finding forums to gather information from energy firms, institutional investors, financial analysts, and other relevant market participants about the company-specific material risks that should be identified and disclosed. The SEC should then institute a system for large publicly listed companies to voluntarily disclose individual corporate climate risks for a fixed period of three to five years. The program should be structured to provide incentives for participation and a credible threat of possible actions for nonparticipation.

- *A major research organization or university should establish a climate risk prediction market.* A well-designed climate prediction market could provide aggregated information for investors and add transparency to how market participants price the probability of climate change outcomes, such as sea-level rise or incidence of heat waves. Charitable foundations and organized financial commodity exchanges, such as the New York Mercantile Exchange and CME Group, should provide financial and technical support for the market.

CLIMATE CHANGE, STORM SURGE, AND THE OIL AND GAS INDUSTRY

Jim Blackburn and Amy Myers Jaffe

Climate change has posed and will continue to pose a threat to the U.S. refining industry. The greenhouse effect, in which the rising level of carbon dioxide and other greenhouse gases in the earth's atmosphere slows the escape of the sun's energy from the lower atmosphere, means an increase in energy in both the atmosphere and the oceans. Although no particular storm can be directly linked to global warming, more energy in the atmosphere and warmer seas increase the probability of a larger number of more severe hurricanes and other extreme weather events. The decline in sea ice both increases the warming of oceans (because dark water absorbs more of the sun's heat than ice, which reflects it) and promotes sea-level rise. Because a high percentage of U.S. refining is coastal, sea-level rise and the potential for more severe storms are both challenges for the U.S. oil and gas industry. Some 120 U.S. oil and gas facilities are located within ten feet of a low tide line (see figure 1).[1]

The U.S. Gulf Coast has one of the country's fastest rates of sea-level rise, due in part to local land subsidence and changes in sea currents. Galveston, Texas, has experienced more than a foot of sea-level rise over the past fifty years; the global average is much lower, according to the National Oceanic and Atmospheric Administration (NOAA). A 2014 study suggests that under a mid-range scenario for warming, the Gulf of Mexico could experience up to four feet of sea-level rise by 2100.[2] Hurricane-associated storm intensity and rainfall rates are projected to increase, according to the 2018 Fourth National Climate Assessment. Heavy rain downpours are also increasing nationally; the largest increases so far are in the Midwest and Northeast, exacerbating regional flooding. Projections of future climate in the United States suggest that the trend toward heavy precipitation events will continue across all regions, including the Southwest.

Figure 1. OIL REFINERY CAPACITY

Bubbles represent oil refineries, sized according to capacity in barrels per day.

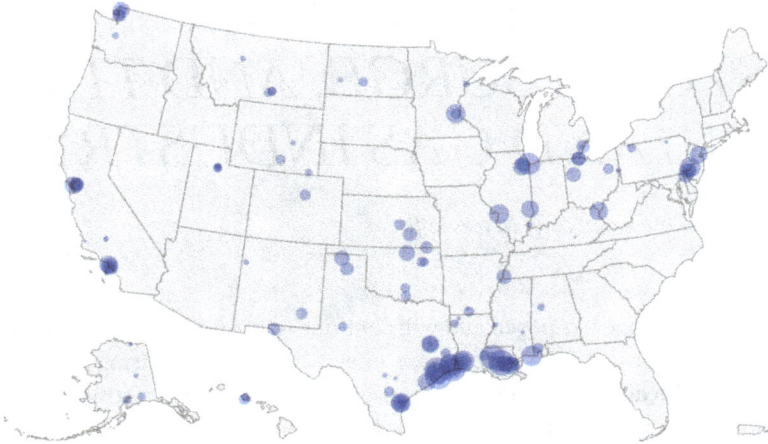

Source: U.S. Energy Information Administration.

As temperatures increase, Gulf Coast storms will continue to worsen. The thirteen most intense hurricanes in the region in the last hundred years have made landfall since 1980, five of those since 2000. No storm of record compares with Hurricane Harvey in terms of rainfall: it dropped close to fifty inches of rain over the central Houston region over four days in 2017. In addition to the category 4 Harvey, two category 5 storms hit in 2017, Hurricanes Maria and Irma. Maria had a particularly large eye, with associated hurricane-force winds extending far beyond the eye wall. Houston has seen a 30 percent increase in its hundred-year rainfall event definitions (a volume of rainfall that has a 1 percent chance of occurring in a given year). No similar statistical update has been conducted on storm-surge flooding.

Northern California also faces unique challenges from climate change. A sea-level rise of between 2.4 and 5.4 inches is projected for the San Francisco Bay Area by 2040. Coastal regions are especially vulnerable to flooding. Some zones are considered vulnerable to wildfires in areas where flammable vegetation and seasonally hot and dry climate have increased risk. Wildfires are expected to increase every few years in areas dominated by grass, shrub, chaparral, and forests, especially when droughts occur and temperatures are high.[3] When wildfires and loss of vegetation are followed by heavy rain, the flooding can weaken

Impact of Climate Risk on the Energy System

hill slopes and lead to landslides, which can put pipelines at risk. In addition, refining requires access to electricity, natural gas, and water; their loss could force a shutdown even if the refinery is not damaged.

REFINING ASSETS AND EXPORT TERMINALS IN THE GULF OF MEXICO

The Gulf Coast region has the single largest concentration of oil-refining capacity in the United States. The region is home to 44 percent of the country's total 18.6 million barrels per day (b/d) of refining capacity. The refineries in the Gulf Coast do not just serve the needs and demands of that region—pipelines also transport petroleum products such as gasoline and heating oil from the Gulf Coast across the nation. The Colonial pipeline delivers petroleum products from the Gulf Coast to the New York area, U.S. Department of Energy Petroleum Administration for Defense District region one (PADD I), making deliveries all along the southern and eastern states. It delivers an average of 2.38 million b/d of gasoline, home heating oil, aviation fuel, and other refined petroleum products. The Plantation pipeline delivers petroleum products from the Gulf Coast to the Washington, DC, area. The TEPPCO pipelines deliver products from the Gulf Coast to the Northeast: the Centennial provides 210,000 b/d and the TE delivers 340,000 b/d. The Explorer pipeline transports 500,000 b/d from Port Arthur, Texas, through the Midwest to Chicago (PADD II). The 9,700-mile Magellan pipeline system is the largest in the United States, extending from the Gulf Coast to the central United States, including Kansas, Minnesota, and Indiana, and carries 500,000 b/d. The Rocky Mountain States (PADD IV) receive products through the Phillips pipeline from the Gulf Coast. PADD V, the West Coast, is the only region that does not heavily rely on Gulf Coast refineries for product supplies.

The Gulf Coast is also home to several major ports that host facilities used in exporting crude oil from the United States to global markets. In Texas, at the end of 2018, Houston-area terminals were shipping about 650,000 b/d of U.S. tight oil, followed by Corpus Christi ports at 570,000 b/d and Beaumont facilities at 485,000 b/d. Often, smaller cargoes are loaded at the Houston ports and then transferred ship-to-ship, in a process called lightering, to larger vessels in the Galveston Offshore Lightering Area or off the coast of Corpus Christi. Small volumes have also begun departing from the Louisiana Offshore Oil Port, which is the only port on the Gulf Coast that can accommodate full loading of very large crude carriers (VLCCs) without lightering transfers from a smaller ship to a VLCC in deeper waters off the coast. All the terminals can handle substantially higher volumes than they currently do, and expansion projects are in the works. Of the seven terminals slated to export U.S. natural gas in the form of liquefied natural gas (LNG) by 2020, six are located on the Gulf Coast. At least two additional LNG export terminals are likely to be built, again on the Gulf Coast. Sabine Pass in Texas is the site of the largest LNG export terminal, boasting the capacity to ship 22.5 million tons per year, followed by Freeport, Texas, at 15 million tons per year.

Although a damaging hurricane does not strike the Gulf Coast every year, the region has been hit by several severe storms in recent years. In 2005, Hurricanes Katrina and Rita made landfall on the Gulf Coast. Hurricane Katrina did so as a category 3 hurricane on August 29, 2005, in southeast Louisiana. It was one of the costliest natural disasters and deadliest hurricanes in U.S. history. Hurricane Rita also made landfall as a category 3 hurricane on September 23, 2005, on the border of Louisiana and Texas. Although not as destructive as Katrina, Rita caused extensive damage along the Gulf Coast. After the two back-to-back storms, about a quarter of all U.S. refining was affected: fourteen refineries, which had a total capacity of more than four million b/d, shut down in late September 2005, and another three facilities operated at partial rates. By November 2005, most of the lost capacity had been restored, though two facilities in Louisiana did not come back online until April 2006. Of the eleven refineries shut down by Katrina, four remained offline sixty days later.

In 2008, Hurricane Gustav was the most destructive hurricane of the season, making landfall as a category 2 in Louisiana on September 1, 2008. It shuttered 2.7 million b/d of refining capacity initially; almost all was restored within a month. Hurricane Ike, the third costliest, made landfall as a category 2 near Galveston, Texas, on September 13, 2008. Ike forced the closure of fifteen Texas refineries and their total capacity of four million b/d. One major plant in Baytown, Texas, owned by ExxonMobil, remained closed for close to a month. But surge waters from Ike missed the Houston-Galveston refining complex as the storm's landfall pushed the major storm-water surge to the east into the mainly undeveloped wetlands and prairies of southeast Texas. After Hurricane Isaac in 2012, a little under one million b/d of U.S. refining went offline temporarily, and one Louisiana refinery was damaged by flooding. By contrast, historic rains during Hurricane Harvey affected operations at

Figure 2. REGIONS MOST VULNERABLE
TO RECURRENCE OF HURRICANE SURGE FLOODING

The areas with greatest risk are also those with the highest concentration of refining capacity, particularly zones C and E.

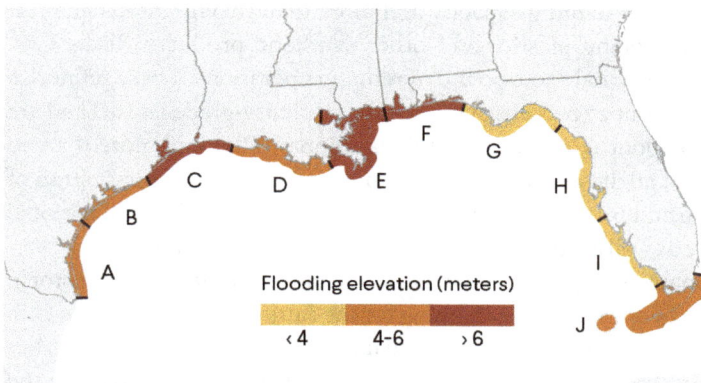

Source: H. F. Needham, B. D. Keim, D. Sathiaraj, and M. Shafer, "2012: Storm Surge Return Periods for the U. S. Gulf Coast."

ten oil refineries and numerous storage facilities in and around Houston. Flooding impeded major refining complexes in the area, including in Beaumont, Deer Park, and Port Arthur, Texas. At their peak, the shutdowns disrupted three million b/d of refined products production.[4]

Based on the historical record of storm surge, the areas of the coast of the U.S. Gulf of Mexico with the highest concentration of refineries also have higher, if not the highest, potential for surge flooding (see figure 2). This prediction encompasses extreme flooding events that have a 1 percent probability of being equaled or exceeded in any given year (a one-hundred-year-flood risk). Of particular concern is zone C—which includes the Houston Ship Channel and Beaumont, Port Arthur, and Texas City in Texas and Lake Charles in Louisiana—and zone E, which includes the Baton Rouge and New Orleans refining complexes in Louisiana. Coastal Louisiana has lost about five thousand square kilometers of wetlands because of river leveeing, dredging navigation canals, and subsurface fluid extraction during oil and gas production and related subsidence. The loss of coastal wetlands and sinking of ground surface lowers the region's ability to absorb dangerous storm surges without destroying highly populated areas. It also increases the chances that sea water will contaminate freshwater sources, damaging fisheries and access to usable water. In southwestern Louisiana, wetlands are projected to be vulnerable to the region's relatively high rate of sea-level rise.[5]

The Houston Ship Channel and Galveston Bay are home to 1.4 million b/d of refining capacity and more than two hundred chemical plants producing plastics and other synthetic products. Texas City adds an additional 830,000 b/d of refining operations. These refineries produce about 27 percent of the nation's military-grade jet fuel and are home to about 13 percent of U.S. gasoline production, more than 30 percent of all diesel fuel, and 15 to 25 percent of the U.S. production of ethylene and propylene. This is an important region for U.S. national security as well as for the Texas and Houston-area economies.

Climate change is at the heart of the question of what size storm needs to be considered in planning for the future of refining on the U.S. Gulf Coast. Recent storms have had unprecedented characteristics, but these conditions are not reflected in flood plain maps prepared by the Federal Emergency Management Agency (FEMA) and are not generally acknowledged within the industrial community. Addressing the discrepancy between perceived and real risk is central to preventing a huge disaster on the Houston Ship Channel.

The west coast of Galveston Bay, where several refineries are located, is extremely vulnerable to storm surge because much of its development is below thirty feet in elevation. To date, the largest surges recorded up the Houston Ship Channel were from Hurricane Ike in 2008 and Hurricane Carla in 1961, and neither storm generated more than thirteen to fourteen feet of surge up the channel. Modeling by Rice University's Severe Storm Prediction, Education, and Evacuation from Disasters (SSPEED) Center reveals that serious inundation of the refinery capacity along this region would occur from a category 3 storm similar to Hurricane Ike but with 15 percent stronger winds. A storm surge of about twenty-five feet could damage or partially damage roughly half of the 4,400 petroleum and hazardous-material storage tanks found along the Houston Ship Channel. This could release more than ninety million gallons of oil and hazardous substances into neighborhoods and the bay, seven times more than from the Exxon Valdez spill and about half that from the Deepwater Horizon spill. Most channel industries and tanks are protected against a surge of roughly fifteen feet. Expanded hurricane wind fields have been observed in more recent storms fueled by hotter water temperatures in the Gulf of Mexico, the Caribbean Sea, and the Atlantic Ocean, raising questions of whether facilities are adequately prepared for potentially larger future storms.

Similarly, refineries in Beaumont and Port Arthur are also vulnerable to the larger storms predicted for the future because the existing seventeen-foot levee at Port Arthur could prove to be ineffective, especially in light of sea-level rise. Surge risk to Corpus Christi is lower, given the narrower continental shelf, but the port will still be at risk for severe storms and rising sea level.

THE SPECIFIC CHALLENGE OF THE BENICIA REFINING COMPLEX IN NORTHERN CALIFORNIA

California is home to seventeen of the nation's refineries, and its market is more disconnected from other U.S. national networks than are those of other regions in the United States. Infrastructure to import refined products by pipeline from other parts of the United States is limited; the state relies mainly on local refineries and seaborne imports from Asia. Its isolation has rendered the California market particularly vulnerable to refining dislocations: over the past several years, gasoline prices have seen a sustained premium that has stemmed from refining accidents and other problems.

The study "Assessing Extreme Weather: Related Vulnerability and Identifying Resilience Options for California's Interdependent Transportation Fuel Sector" concluded that most of the state's refinery assets are located near waterways and exposed to coastal flooding.[6] In Northern California, five facilities are located between Benicia, Martinez, and Richmond in the San Francisco Bay Area. The Sacramento-San Joaquin Delta is likely to be extensively flooded as sea-level rise increases, and "terminals, docks, and refineries have higher proportions exposed than other kinds of fuel system assets." The study also found that although terminals and refineries are less exposed to large wildfires than other kinds of assets are, the threat "still exists and is projected to persist." The study concluded that refined-product pipelines are at the highest risk of wildfire-related disruptions, especially when rainstorms that follow fires produce flash floods, slope failures, and debris flows that can destroy infrastructure. Flooding exposure in Northern California around the Concord-Martinez-Richmond complex and wildfire exposure in the Reno-Richmond-Sierras complex were identified as potentially affecting operations of Kinder Morgan's common carrier pipeline system for petroleum products.

POLICY RECOMMENDATIONS

Policymakers have increasingly focused on infrastructure that needs to be constructed to protect facilities. In Houston, experts have proposed an option for Galveston Bay that involves the construction of a "coastal spine" levee and seawall, which would have an elevation of seventeen feet at the coast. The U.S. Army Corps of Engineers proposed a back-side levee for the city of Galveston and two gate structures for the west side of Galveston Bay. The project has encountered criticism not only for its large price tag of $14 billion to $20 billion but also because it would restrict tidal exchange within a productive natural estuary. The project would not be completed until 2035 at the earliest and would not protect the Houston Ship Channel or Texas City refinery complexes from the major storms that are predicted for the future or from the associated sea-level rise. Methodologies used by the Corps of Engineers tend to approximate higher-risk, higher-intensity storms by measuring past storms rather than identify the future risk of greater storms created by the combination of climate change and rising sea levels. As new kinds of storms emerge, historical metrics are not as useful for assessing future infrastructure needs in light of a changing climate.

An alternative plan, the Galveston Bay Park plan, designed by the SSPEED Center, proposes a twenty-five-foot dike that originates in Chambers County, proceeds westward to the Houston Ship Channel, and eventually connects to the Texas City levee system, which will also be raised to a twenty-five-foot elevation. A navigation gate that regulates water levels to enable shipping would also be included, as in the coastal spine project. The Galveston Bay Park plan would also include a backside levee for the city of Galveston and elevated roads. Total project costs are estimated at $3 billion to $6 billion, with possible cost-sharing opportunities with the Port of Houston, which plans to widen the ship channel. Integrating hurricane surge planning with ongoing plans to widen the Houston Ship Channel and using dredged material to build the barrier as part of the dredged-material disposal plan could save time and expense. Gate structures would be added as a second stage. Other improvements such as elevated highways would be built in conjunction with federal and state highway projects, and additional levees could be constructed by the Army Corps of Engineers.

Underlying these plans and philosophies about flood protection is a fundamental issue. The climate is changing, and older planning concepts and methodologies are inadequate to address this future risk. New planning and infrastructure evaluation methodologies that specifically incorporate climate change are a fundamental priority in protecting critical fuel infrastructure for the future. This problem can be resolved not by looking to the past but by looking forward and anticipating the changing climate. Otherwise, big storms such as Harvey will always catch an area off guard.

In addition to major initiatives such as constructing seawalls and dikes, building standards for storage tanks should be revisited and updated to reflect projections of future risks. The refining industry has been relocating critical equipment to elevated areas, but it will need to reevaluate whether current standards will be sufficient as sea-level rise and larger storm surges occur. Hurricane Harvey highlighted the need for a major reassessment of the planning procedures of entities such as the Army Corps of Engineers. Climate change should be addressed directly to plan for the potential storms of the future. In this regard, considering staged protection is useful: some protection is required now; greater protection could be required in twenty to forty years. That reality would suggest building in increments over time rather than attempting to solve all problems with the construction of single, massive structures.

Today's storage tanks are considered highly vulnerable to severe storms in that they are designed to withstand winds but not necessarily rising water. Better processes and storm preparation procedures are needed, including emptying tanks (or perhaps filling them per weight) in anticipation of storms. Containment berms for spill events can become traps for rising water—not high enough to keep the water out and too high to let the water leave. Electrical systems should be elevated well above ground level, as should pumps, compressors, and generators. Facilities should have flooding management plans and operational provisions for flooding that take into account the power of a coastal surge and consider the costs and liabilities of tank failure and leakage of toxic materials that are inside equipment in refineries and petrochemical plants that would be catastrophic to adjacent communities and the environment.

In California, refineries and pipeline operators should revisit how they manage fire-prone vegetation near their facilities to ensure that they are aligned with the latest data on year-round fire risks. Wood materials need to be upgraded to concrete and steel. California should also consider how to promote more redundancy in fuel transportation to markets so supply routes do not depend on a single railroad or highway for trucking. Further study on preparedness for fuel disruptions is likely to show a greater need for pre-positioned fuel inventories, especially for first responders. The state should consider regulations and incentives to encourage holding higher fuel inventories at in-state locations that are less subject to flooding, rain, and fire.

Regulators in states along the Gulf Coast should require refiners and the petrochemical industry to compile and maintain a current inventory of hazardous chemical volumes that are stored at or near their facilities. Credit rating agencies should consider these inventory stockpiles and the history of environmental compliance and safety records, including spill records, in accounting for risks and performance of Gulf Coast refining and petrochemical companies. Federal law requires companies to report spills of hazardous chemicals, but in the aftermath of Hurricane Harvey, investigations of the many incidents have been spotty.[7]

Texas has not yet tackled how damages from many of the spills will be ameliorated and who will pay the related costs. Given that similar catastrophes could become more frequent in the future, federal and state authorities should consider creating permanent disaster-response funding, separate from FEMA assistance. This funding could be used to settle claims from toxic releases and fund restoration of soil, waterways, and other harmed ecosystems, as well as cover state and local response costs and individual compensation. A response fund could include a structured claims process and a court-supervised settlement process. Precedents such as the Deepwater Horizon Oil Spill Trust and the 1980s Superfund program can serve as a starting point for studying the proper structure for such disaster-response programs. Public policy debate needs to weigh the appropriate cost sharing from industry and public funding. Future policy design will need to ensure that private insurance markets can continue to function adequately and that market failures, like those looming in light of the bankruptcy of California utility PG&E, can be anticipated and mitigated.

CONCLUSION

The forty-seven-plus inches of rain over four days from Hurricane Harvey should have convinced Gulf Coast refining and petrochemical operators of the need to recalibrate rainfall expectations and reconsider adaptation plans to protect the nation's refining capacity in light of climate change. Future consequences of hurricane surge in light of sea-level rise will be a greater threat to Gulf Coast refining than in the past. Similarly, heavy rains, sea-level rise, and year-round fire risk in California could create new challenges to local fuel infrastructure. Another risk is cascading effects throughout the refining system, especially given that refineries are unable to operate or distribute any existing fuel output stored in tanks without reliable electricity supply.

The need for detailed, location-specific information about changing patterns of weather-related risks will only increase. Companies and political leaders alike will need to incorporate updated analysis into forward-planning and adaptation activities. The task is daunting and requires collaboration at all levels of government, citizen action, and industry. Uncertainty is not an invitation to inaction. Rather, a healthy debate about the risks facing the U.S. fuel system and about the various options that can mitigate those risks is imperative to U.S. energy and national security, as well as the welfare of communities near energy infrastructure.

WATER-RELATED RISKS AND IMPACTS ON THE U.S. ENERGY SYSTEM

Christina Copeland and Sara Law[*]

Water and energy are closely connected. Water, particularly fresh water, is important in most forms of energy production and electricity generation. The Intergovernmental Panel on Climate Change (IPCC) has found that rising greenhouse gas accumulation in the atmosphere threatens the reliable availability of fresh water.[1] Droughts and lost snow cover could shrink renewable surface water volume in rivers and reservoirs. With lower rainfall, high water use could deplete ground-water in dry, subtropical regions. In arid areas, climate change could increase the frequency and severity of droughts. Scientists at Columbia University's Lamont-Doherty Earth Observatory have found evidence that over the past two centuries droughts in the western United States have worsened because of climate change.[2]

Significant changes in water quantity and quality are evident across the United States. Variable precipitation and rising temperatures are intensifying droughts, increasing heavy downpours, and reducing snowpack. Reduced snowpack, in turn, worsens drought effects because less water is stored in the form of immobile snow that can be gradually released as seasons change to replenish water supplies during the hotter, drier months. Thus, climate change effects are creating a significant discrepancy between human water withdrawals and natural reservoir replenishment. Groundwater depletion is exacerbating drought risk, especially in California and much of the Southwest.[3] Surface water quality can also be expected to decline as water temperature increases, and as erosion and runoff increase as a result of more frequent high-intensity

[*] The authors work for CDP Worldwide. They would like to acknowledge Zoya Abdul-lah and Benjamin Silliman for their contributions to this essay.

rainfall and recurrent flooding from sea-level rise. During Hurricane Harvey in Texas, thirteen of forty-one Superfund sites in Texas were affected by the flooding. Superfund sites are areas so polluted from severe hazardous waste contamination that the U.S. Environmental Protection Agency identified them as a national priority site for cleanup. Additionally, nearly five hundred chemical plants, ten refineries, and 6,670 miles of pipelines were in the path of the hurricane.[4]

As climate change threatens to change water patterns around the country, energy security could be put at risk. The U.S. energy industry uses substantial volumes of water in its everyday operations. Almost all forms of energy production rely on a stable supply of good-quality water. Of the water used by the energy industry, 76 percent is fresh water.[5] Reliance on fresh water puts the energy industry in competition with other industries, such as agriculture, as well as with human consumption. Energy companies are already suffering from costly operational stoppages and financial losses related to severe water shortages that were not properly anticipated or managed. Water scarcity can raise energy companies' operating, financing, and insurance costs, as well as disrupt supply chains, constrain growth, damage brand names, and, if long lasting, strand assets.

Water issues have already emerged as a risk to energy production in parts of the United States. In 2013, Antero Resources Corporation proposed a pipeline to carry water from the Ohio River to hydraulic fracturing (fracking) sites in the Marcellus and Utica shales when state authorities in Ohio curtailed water withdrawals from local rivers during the 2012 drought in the region.[6] California's hydroelectric power production was cut in half in the mid-2010s, and electricity production at various U.S. coal plants has been hampered by water problems in recent years.[7] Drought also affected nuclear plant operations in the Southeast, Illinois, and Minnesota in 2006 and 2007. In July 2012, heat waves and drought forced nuclear plants in Ohio and Vermont to slow output.

That said, much of the attention on water use in U.S. energy production has focused on the exploitation of unconventional resources in Ohio, Pennsylvania, and Texas. Energy and research consultants at Wood Mackenzie projected in 2013 that more than half of the shale and tight natural gas reserves in the United States "are located in medium to extremely high baseline water stress areas, where competition is high with other local water users and concerns over water quality exist."[8] U.S. water withdrawals for fracking rose from 5,600 barrels of water per oil well in 2008 to more than 128,000 barrels by 2014.[9]

To put these statistics in perspective, water use for oil and gas drilling represents only a small portion of the water use for energy in the United States, and energy use is only 3 percent of total U.S. water consumption. Yet water is a critical component of electricity production: electric utilities withdraw large amounts of water for cooling purposes and then return it to the water cycle after use. According to the U.S. Geological Survey, thermoelectric power accounts for nearly 45 percent of fresh water withdrawn from natural sources in the United States (and 3 percent of total water consumption).[10] Some 70 percent of current U.S. electricity comes from power plants that require water for cooling. Less carbon-intensive power sources, such as biofuels, nuclear power, and, of course, hydropower, all require significant amounts of water. Even solar photovoltaics, one of the least water-intensive sources of power, requires water for equipment maintenance.

Several factors will influence the water requirements of the U.S. electricity grid moving forward, including fuel consumption patterns, cooling technologies in use, environmental regulations, and ambient climate conditions. The water requirements for generating electricity are related to the kinds of generation plants that make up the grid. Different kinds of electricity-generation plants have varying water requirements for their operations. For example, thermal coal plants and nuclear energy are relatively water-intensive, whereas solar and wind energy do not require much water. Companies need to consider future water availability when deciding which kinds of generation facilities to build and what equipment and operational practices they will use to lessen water requirements.

ANALYSIS OF U.S. ENERGY COMPANIES' WATER RISK DISCLOSURES

How electric utilities companies manage climatic water risk will affect the amount of water needed to maintain U.S. electricity supply. Water users need to clearly understand the risks of potential water supply problems and the methods that can be deployed to appropriately manage water resources. This would mitigate risks to business and protect water ecosystems.

Increasingly, energy companies are assessing whether their operations could be interrupted by climate-induced water scarcity. The type of fuel used to generate heat in plants can determine the plant's water use. Power plant cooling typically drives the largest need for water in thermoelectric plants. Nuclear power plants typically operate at lower thermal efficiencies and require more steam per unit of power generated, increasing their water requirement. The replacement of coal-fired plants is likely to lower the water intensity of the U.S. electricity grid over time.[11] Greater reliance on wind and solar energy, as well as on smaller, distributed micro-grids, can relieve some water requirements now required for large-scale thermal coal plants.

Shareholders are requesting that companies disclose their water-related risks and report steps they will take to reduce vulnerability to water-related disruptions to their operations. One recent example is the Arizona Public Service Company (APS, a subsidiary of Pinnacle West), which disclosed to investors its plans to retire 767 megawatts of coal-fired generation by 2025, a move projected to reduce water consumption at the Cholla Power Plant to less than 10 percent of current consumption. APS has retired coal units totaling 820 megawatts since 2013, thus reducing water consumption by approximately 20 percent. Because trillions of dollars' worth of assets are set to be at risk from water insecurity, investors are more focused than ever on leaders and laggards in the sustainability transition. Information is fundamental to their decisions.

Disclosures by U.S. energy companies to the nongovernmental organization CDP Worldwide reveal large variation among energy companies in how they are managing water risk. CDP analyzed the water disclosures of twenty large representative U.S. energy companies. The primary activities of these companies included coal mining, oil and gas extraction, oil and gas midstream, and thermal power. The disclosures reflect the companies' responses to CDP questionnaires created with the backing of more than 525 institutional investors with assets of

$96 trillion. CDP's questionnaire focuses on water stewardship, looking at realized effects and expected risks as well as a company's responses to and plans for these effects and risks. Questions included reporting on whether the company "experienced any detrimental water-related impacts" in the reporting year (question W2.1), the number of facilities exposed to current or future water risks "with the potential to have a substantive financial or strategic impact on [their] business," and explanations of what proportion of their company-wide facilities this represented (question W4.1b). Companies were also asked to provide details of the identified risks (effects on their direct operations, including potential financial costs) and their responses to those risks (question W4.2). CDP has a scoring system based on the answers to the questionnaire. More details on individual company scores and the metrics used that encompass the one hundred to three hundred data points covered in each company's 2018 responses to the questionnaire are published on CDP's website.[12]

In its analysis of the companies' disclosures of effects, CDP found that these twenty representative energy companies have already experienced detrimental water-related disruptions to their operations. Water scarcity caused the twenty companies to experience $1.8 billion in aggregate revenue losses over 2017. In the 2018 disclosure report, companies cited hurricanes, drought, wildfires, and sea-level rise as some of their past or anticipated water-related challenges. Multiple companies also cited regulatory compliance and uncertainty as ongoing risks.

The companies reported 271 facilities as being exposed to water risks that could have a substantive financial or strategic effect on business operations. Eight companies reported that their exposed facilities represented up to 25 percent of company-wide facilities, five reported exposure of 26 to 75 percent, and six reported exposure of 76 to 100 percent of company-wide facilities.

A total of ninety-two risks were reported, fifty-one physical and thirty-eight regulatory. (A small number of reputational or market risks were also identified.) The most commonly reported risk driver was drought, followed by regulation of discharge quality or volumes. More than a third of the risks reported are already happening or will take place within the year. The total potential financial cost of those risks was in excess of $30 billion. However, CDP considers that the ultimate financial effect could be much higher because some of the reported risks did not contain sufficient information on expected financial implications.

The most commonly reported potential consequences of these risks were an increase in operating costs and a reduction or disruption in production capacity. For example, Exelon Corporation reported that at its Oyster Creek facility in Barnegat Bay, New Jersey, the cost to meet regulations to reduce the risk of ecosystem vulnerability by installing closed-cycle cooling towers would have been more than $800 million over the remaining twenty-nine years of the life of the plant; it closed the facility instead.

Six companies reported one or more of the following physical implications: disruption of sales, effect on company assets, increased operating costs, reduced revenues from lower sales or output, reduction or disruption in production capacity, up-front costs to adopt or deploy new practices and processes, and other conditions.

Hurricane Harvey featured widely in the disclosures of several companies. For example, American Electric Power Company reported that flooding due to the hurricane caused oil-filled electrical equipment to fail, which then also led to oil spills and releases to the environment. The cleanup of all Harvey-related spills cost the company approximately $448,000. Occidental Petroleum Corporation reported that flooding effects from Harvey resulted in its suspending operations at some facilities. In its reporting to CDP, the company estimated that "realized losses" attributed to the hurricane included a pre-tax income reduction of approximately $70 million.

For Duke Energy Corporation, Hurricane Irma was the most destructive storm to its Florida service area: almost 75 percent of Florida customers lost power for up to eight days. Duke lost $513 million, but the company reported it to CDP as "eligible for recovery," expecting to offset the loss with federal tax savings because of the storm costs. Duke plans to invest $3.4 billion over the next ten years to strengthen its Florida energy grid, including by moving about 1,250 miles of its most outage-prone overhead power lines underground.

Five companies reported regulatory factors as affecting their operations and capital spending. American Electric Power Company reported that new rulemaking on discharge of pollutants in flue-gas desulfurization wastewater and discharge restrictions on bottom ash transport water could increase compliance costs. CMS Energy Corporation reported that new environmental regulations on discharges of cooling water could require the company to make expensive modifications at two of its facilities. DTE Energy Company's statement to CDP also underscores its view on regulatory risk. It wrote that new water discharge regulations "imposed a significant financial burden to the company, and were one of many contributing factors to several plant closures." DTE and other companies reported effects related to compliance with the coal combustion residuals rule, which regulates coal ash storage sites to prevent environmental harm, such as to groundwater sources.

Too little water was also reported as a risk. Sempra Energy reported that drought conditions in the western United States "increase the risk of catastrophic wildfires," which it added "could place our electric and natural gas infrastructure in jeopardy." It described its potential liabilities for massive damages caused by fires linked to its equipment failures as follows: "If overhead power lines owned by our business unit SDGE [San Diego Gas and Electric] are implicated in wildfires, as was the case in 2007, it represents further financial risk, through increases in insurance and litigation costs." SDGE proposed passing on $379 million in expenses from three 2007 wildfires in San Diego county to consumers, but so far regulators have rejected that remedy. The company's 2018 wildfire liabilities could be higher. Another factor is that wildfires can leave areas prone to flooding during the rainy season, further endangering underground infrastructure.

Sempra Energy also evaluated the effect of rising sea levels on electric and natural gas infrastructure on its subsidiaries Southern California Gas Company and SDGE. It reported a potential financial burden of $25 billion, representing "potential costs to customers as a result of indirect impacts of coastal climate change hazards on the economy and social fabric of the San Diego region, under an extreme scenario by the end of the century."

Beleaguered California electric utility PG&E Corporation, which is in bankruptcy and faces billions of dollars in potential wildfire liabilities, reported to CDP that its facilities also face high risk of flooding related to sea-level rise and storm surge: "There is the risk of levee erosion or failure, putting assets at risk. PG&E also faces the risk of damage to substations and other gas and electric infrastructure." It partnered with the Center for Catastrophic Risk Management at the University of California, Berkeley, to better understand how its gas transmission infrastructure could be affected under future sea-level rise coupled with a storm surge event. Based on a preliminary review of a worst-case scenario of 1.4 meters of sea-level rise coupled with a major storm event, PG&E estimated that mitigation efforts would cost $4 million to $7 million annually.[13]

ADDRESSING WATER SECURITY RISKS

The business model that energy companies adopt today will have ramifications for decades. Companies with a greater understanding of water insecurity will be better positioned to adapt their operations to withstand the pressures of freshwater scarcity and thus improve energy security. Those companies could decide on a mix of fuels available in a particular geography that would minimize water use. These actions would position them to achieve the transition to net-zero emissions and reduce the risk of climatic water shortages to the U.S. economy. Thirteen of the twenty companies analyzed stated that they use climate-related scenario analysis to inform business strategy. Of the thirteen, seven provided details of that analysis, but only three reported using established scenarios such as 2 degrees Celsius (2DS), International Energy Agency (IEA) 450, and the IEA Sustainable Development Scenario.

CDP's analysis leads to several suggestions as to how U.S. energy companies can improve their response to these costly risks and consequences. A faster push to renewable energy could ease exposure to water constraints. Renewable energy requires less water for operations. Hastening the energy sector's shift to renewables will provide cleaner energy and support a society-wide shift to a low-carbon economy. The deployment of new, large-scale electricity generation powered by renewable sources is also allowing researchers to examine how this form of power fares when exposed to extreme storms. Many current water-related effects and future risks reported by U.S. electricity companies were associated with extreme weather, specifically hurricanes. Renewable energy facilities have consistently outperformed fossil fuel plants in the wake of destructive hurricanes. For example, after Hurricane Matthew in 2016, North Carolina worked to rebuild a system with more solar and wind energy. Renewable power now accounts for 10 percent of its total electricity production.[14] After Hurricane Florence in 2018, thousands of residents in the Carolinas whose electricity comes from coal-fired utilities were without power for several weeks. By contrast, solar installations were reported to be up and running the day after the storm.[15]

Several technologies can be installed relatively quickly to help traditional power-generating units reduce water use, which could have a more immediate benefit to a company's exposure to water insecurity. Hybrid wet and dry cooling systems and closed-loop water cooling systems could reduce water withdrawal in areas where water is becoming scarcer. (These technologies, however, could increase water consumption by increasing evaporation and slightly lowering the thermodynamic efficiency of the power plant.) In areas where water is more readily available, such as the eastern United States, cooling systems that return water at the same rate it is used could reduce the water consumption rate and improve the efficiency of the cooling system. Reclaimed wastewater can also be used as a supplemental source for cooling water if the infrastructure is created to do so. On the extraction side, some oil and gas companies are already using artificial intelligence and automation to deliver water for operations in the exact amount and timing that is needed, thus reducing waste.

Exploring collaborative partnerships among utilities, suppliers, government agencies, and communities is important to ensure that all parties understand the shared stake in water security risks and can take coordinated actions to ameliorate those risks. Among the twenty companies analyzed, local communities, regulators, and river basin management authorities were the most commonly mentioned in companies' water-related risk assessments. This is a promising sign of progress in addressing shared water challenges, especially if companies move to enhance collaborations at the local level. When companies undertake a water-related risk assessment, they should consider not only the frequency with which they update that assessment but also at what geographical scale they are considering their risks—and thus which groups they should work with. For example, through collaboration with communities, industry groups, and regulatory agencies, a company can monitor and address water supply risks at the company, aquifer, or watershed level—that is, consider the broader system of connected streams and rivers that flow into the same outlet, such as a large river, lake, or ocean. Because risks to site-level water access manifest at this broader regional level and large withdrawals from any one source can affect water availability in other connected bodies of water, including this wider scale in corporate monitoring and disclosure is vital.

Existing technologies and techniques can help existing or future energy installations lower their water risk exposure. However, these options should be evaluated on a site-specific basis, so that each plant or extraction site can be studied to determine the most effective mechanisms to reduce water use for that operation. No one option will fit all conditions. Adequate disclosure and planning systems can help guide this process. Federal and state governments should further standardize water risk disclosure requirements, which would improve understanding of water needs and improve the availability of data on adaptation to evolving water availability.

CONCLUSION

Access to water (particularly fresh water) is critical to secure energy production today and in the future. Climate change will harm this interrelationship in ways that the government, communities, and corporations need to manage better.

CDP's analysis of U.S.-headquartered energy companies that made disclosures in 2018 in response to an investor-driven information request found that the twenty companies reported current water-related issues costing nearly $2 billion annually, and future risks with a potential financial cost of more than $30 billion. The companies' disclosures also told the story of how extreme weather affects their business, including service interruptions and unrecoverable losses, strategic decisions to retire plants deemed vulnerable, and unstable water supply or new regulatory requirements.

To mitigate water-related risks, companies need to better understand their sensitivity to future water shortages and swiftly adapt. One option for utilities is to replace water-intensive thermal generation plants with renewable energy that requires less water to produce electricity. As more renewables are deployed in the United States to lower carbon emissions, the net result could also help reduce the water needs that could result from any rise in electricity demand.

CLIMATE CHANGE IMPACTS ON CRITICAL U.S. ENERGY INFRASTRUCTURE

Joan M. Ogden

The U.S. energy system is already feeling the effects of extreme weather events and changing climate. The United States' 2018 Fourth National Climate Assessment notes that the United States will be threatened by more frequent and longer-lasting energy outages and fuel supply imbalances that can create "cascading effects on other critical sectors."[1]

Observed climate effects include rise in air and water average temperatures, more extreme temperatures, sea-level rise, increasing intensity and frequency of storms and storm surges, increasing intensity and frequency of flooding, more frequent and severe wildfires, changes in precipitation patterns, drought, decreasing snowpack, and decreasing water availability.

These climate change consequences can affect every part of the energy system, from production through end use, resulting in multiple problems:

- less efficient electric generation, transmission, and distribution due to higher temperature

- wildfire damage

- flood damage

- storm damage

- increased risk of physical damage and disruption to power and fuel facilities

- disruption of rail and barge transport of crude oil and other petroleum products

- increases in air conditioning and natural gas demand in summer

These energy system effects are being felt throughout the United States, though they vary by region (see figure 1).[2]

Rising air and water temperatures throughout the United States reduce the efficiency of electric transmission lines and thermoelectric power plants that make electricity from heat, such as nuclear power plants and combustion turbines fueled with natural gas or coal. In thermoelectric power plants that require water for cooling, hotter water temperatures (sometimes coupled with water shortages) have led to shutdowns or curtailments.

Wildfires threaten the electric grid throughout the western United States with direct damage to the system and prolonged power outages. The past few years have seen billion-dollar wildfires in Alaska, California, Montana, the Pacific Northwest, and the Southeast.[3]

Droughts have become increasingly frequent and severe in the West and South and have also caused major damage in the East and Midwest. Consequences for the energy system include decreased availability of hydropower; decreased crop yields, which could limit biofuel availability; and scarcity of cooling water for power plants.[4]

Rising sea levels and the threat of storm surges affect all coastal regions of the United States and all aspects of the energy system, including electric generation, transmission, distribution, and fuel production, refining, and distribution. The threat is amplified along the Eastern Seaboard and Gulf Coast by the rising frequency of hurricanes and superstorms, and in the West by the increasing frequency of atmospheric rivers (storms that result in extreme downpours and floods). Extreme rainfall events are leading to flooding inland, as well as along the coast, with major damages occurring in the interior states of the

Number of billion-dollar weather and climate disasters (CPI-adjusted), 1980 to 2019

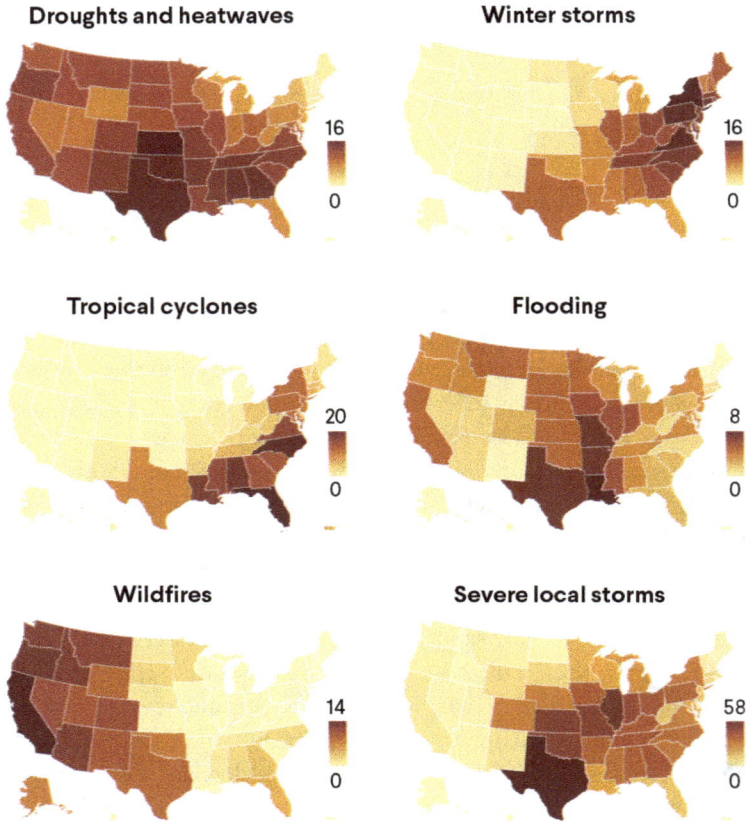

Droughts and heatwaves — Winter storms — Tropical cyclones — Flooding — Wildfires — Severe local storms

Source: National Oceanic and Atmospheric Administration.

Midwest and East, as well as the West. Severe local storms and tornado outbreaks have become more common in the Midwest and Southeast. According to the Fourth National Climate Assessment, multiple climate effects often occur in a given region, where they can interact or compound, amplifying the damage to the energy system:

- Higher air and water temperatures could contribute to both an increase in electricity demand and a decrease in electricity supply.

- The effects of sea-level rise could be exacerbated by more severe storms and coastal erosion, causing flooding across a larger area. Storms can also damage natural features, such as wetlands, and manmade structures, such as sea walls, that help protect coastal infrastructure from sea level rise and storm surges.

- Both warmer temperatures and drought heighten the risk of wildfires, which—alone or in combination—could ultimately limit the amount of electricity that can be generated and transmitted during times of peak demand.[5]

The upward trend in climate-related extreme weather events since 1980 has been steep, increasing fourfold from 2007 to 2016 compared to 1980 to 1989. Damage costs have grown by a factor of almost three; severe storms and hurricanes are particularly costly.[6] Overall, related damage costs were estimated at about $85 billion in 2018, roughly twice the average annual damage cost between 2007 and 2016.[7] Different regions of the country are more strongly affected by different types of extreme weather and climate events. Wildfires are predominant in the West, and winter storms and severe local storms and tornadoes are more likely in the East and Midwest (see figure 1). Tropical hurricanes are found along the East Coast and Gulf Coast.

Weather is the largest cause of major grid outages in the United States.[8] In 2012, Hurricane Sandy inflicted about $2 billion of damage on New York utility Consolidated Edison Company (Con Edison) and New Jersey utility Public Service Electric & Gas. Rebuilding Puerto Rico's electricity grid after Hurricanes Maria and Irma (2017) was estimated to cost $17 billion. Costs to the energy industry from Hurricanes Katrina and Rita (2005) were estimated at $15 billion. In the wake of the devastating 2018 Camp Fire in Northern California, insurers estimate damage costs at $7.5 billion to $10 billion.[9] A federal judge overseeing PG&E Corporation's legal troubles recently proposed requiring the utility to inspect and rate the safety of its entire network, to greatly reduce the chances that its equipment could ignite additional fires. PG&E estimated that the measures needed to make its grid system resilient against future wildfires would cost upward of $75 billion to $150 billion and require a fivefold increase in electricity rates. PG&E has filed for bankruptcy protection.[10]

In the past, utilities in California have accessed both direct insurers and reinsurers to cover potential liabilities. As insurance premiums rise precipitously in the state, private insurance markets could fail.

The California state government recently established a commission to consider how to insure against mounting climate risks. To date, there has been little progress on how to solve the problem of financing catastrophic outcomes, such as wildfire damages to the electricity grid.

The interdependent nature of different aspects of the energy system means that many climatic events can have a cascading effect, where a shutdown of one element of the system can lead to failures in other parts of the system. For example, when the grid fails it could become impossible to pump gasoline or access the internet. Widespread loss of power can potentially shut down the entire fuel supply chain: oil pumping, pipelines, oil refineries, and retail fuel distribution at gasoline stations. The reliability of electricity undergirds the entire energy system (see figure 2). Cutoffs in energy can also disrupt other vital emergency services across the United States.[11] Telecommunications, trains and subways, data clouds, and medical systems are more interconnected than ever to the grid. Police, firefighters, and other emergency workers need access to fuel. Because many of these public services rely on the electricity grid, resilience of the power and fuel sector against climate change is crucial to future adaptation design.

CLIMATE RISK AND THE ENERGY TRANSITION

Addressing climate change means today's energy system needs to transition to lower or near-zero carbon emissions. This transition could be designed in a manner that improves the energy system's resilience to worsening climatic events.

The U.S. electricity grid is evolving from a twentieth-century, fossil fuel–based, one-way supply chain to a new, interactive, twenty-first-century smart grid characterized by low-carbon distributed storage. This new system includes a high proportion of variable renewables such as solar and wind that are harder to dispatch without electricity storage because the highest availability of wind and solar energy does not always match peak demand, which tends to be highest in the early evening. To capture this energy, it could be possible to use an electric car's battery as storage via vehicle-to-grid technology, use bulk electric storage or distributed generation, or better manage demand.[12]

New emerging digital technologies entering the transport sector, such as electrification, automation, and share mobility, are linking transport to the electric sector in new ways. Artificial intelligence technologies are being adapted to enable autonomous, connected

Figure 2. EXAMPLES OF CRITICAL INFRASTRUCTURE INTERDEPENDENCIES

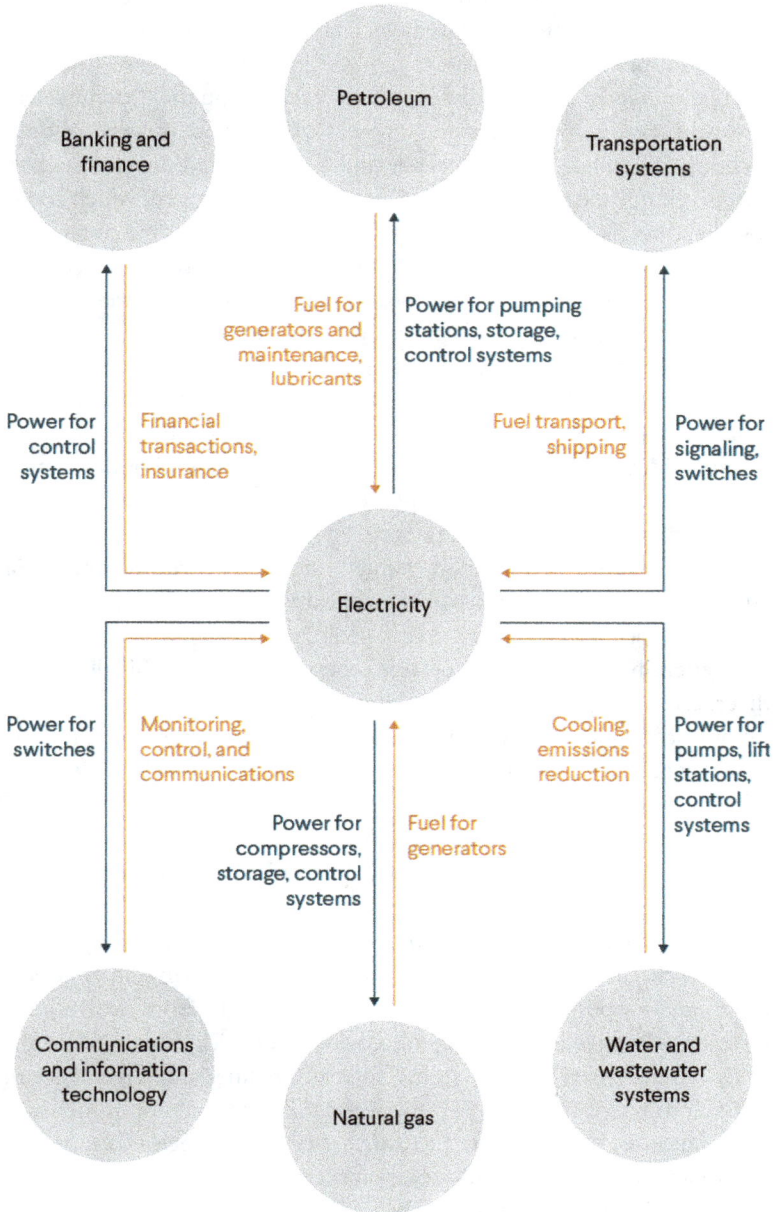

Banking and finance

Petroleum

Transportation systems

Fuel for generators and maintenance, lubricants

Power for pumping stations, storage, control systems

Power for control systems

Financial transactions, insurance

Fuel transport, shipping

Power for signaling, switches

Electricity

Power for switches

Monitoring, control, and communications

Cooling, emissions reduction

Power for pumps, lift stations, control systems

Power for compressors, storage, control systems

Fuel for generators

Communications and information technology

Natural gas

Water and wastewater systems

Source: Fourth National Climate Assessment.

vehicles with the promise of reducing traffic congestion and improving safety.[13] As influential markets transition to lower-carbon targets, new types of vehicles and fuels, including biofuels and both plug-in and hydrogen fuel cell–operated electric vehicles (EVs), will likely be widely adopted by mid-century.[14]

The transport sector will depend increasingly on the evolving electric grid, through EV charging, power-to-gas, and power-to-liquid technologies, wherein excess renewable solar or wind power could be used to charge battery EVs or make electrolytic hydrogen, which could be used directly in fuel cell vehicles or as a feedstock for making transportation fuels. Widespread use of new lower-carbon transport fuels will require different infrastructure strategies. This could mean using or adapting existing infrastructure: implementing smart EV charging technologies into the grid, developing "drop-in" biofuels such as bio-diesel or renewable methane that can use existing fuel pipelines and storage, or even blending modest fractions of hydrogen into the natural gas grid. Eventually, the transformation could require building new dedicated infrastructure for zero-carbon hydrogen or biofuels.

A future near-zero-carbon energy system could also require new infrastructure for carbon management, including carbon capture via chemical processing or atmospheric capture, and carbon sequestration (via CO_2 pipelines and underground geological storage).

Generally speaking, renewable energy is geographically more diverse, reducing the chances of a high proportion of supplies being cut off at once. Solar energy is also easier to restore or replace than thermal power stations or nuclear plants. By contrast, much of the U.S. oil refining and distribution system is concentrated in certain parts of the United States that are highly vulnerable to disruption by extreme weather events.[15] While it could seem like a transition to electric vehicles could increase vulnerability of passenger travel to electric power outages, in fact, electricity is also essential at almost every stage of the gasoline fuel distribution system. Electricity is needed in refinery operations and for pumping gasoline into delivery trucks from distribution terminals. Retail gasoline pumps also need electricity to operate. Refineries and fuel distribution systems, such as pipelines and terminals, are difficult to repair if they suffer physical damage in a storm, sometimes taking months or even years to restore. Several refineries damaged in Hurricane Rita, for example, took over six months to bring back online. Pipelines disrupted by floods can be similarly time consuming to repair. EV chargers could be easier and faster to restore or replace when damaged. EVs can also be charged at home once electricity is back online.

After multiple hurricanes slammed into South Florida in 2004 and 2005, residents were left not only without electricity but also without access to gasoline. In October 2005, Hurricane Wilma caused widespread power outages and forced hundreds of gas stations to close. According to a legislative staff analysis of the bill that became the Florida Alternative Power Source Law, service plazas on a portion of the Florida Turnpike experienced traffic backups of more than five miles, with motorists waiting in line between three and five hours to buy gasoline.

In 2006, Florida lawmakers enacted the law, requiring filling stations within half a mile of an evacuation route or interstate highway to install transfer switches that allow them to switch to generator power in case of an emergency so they can keep pumping gas. Large oil companies with more than ten stations in one county—such as ExxonMobil and Shell—are required to have portable generators available within twenty-four hours. (Louisiana enacted a similar law in 2007.)

Installing the switch is required by law, but it is up to the store owner to supply a generator. Some small station owners buy generators as a form of insurance. A spokeswoman for ExxonMobil told the *Palm Beach Post* that their generators are kept in "strategic locations" but was unable to say where they are and how long it could take for a generator to be transported to a station. Shell-branded stations plan to be able to receive a generator in affected locations within twenty-four hours after a disaster.[16]

In October 2013, New York Governor Andrew M. Cuomo announced that 258 gas stations in the downstate New York area were in the process of installing backup power capacity. He had recently signed a law establishing the strongest protections in the nation to ensure that critical gas stations have backup power capacity, making it possible to avoid long lines and restore normalcy as quickly as possible after a major storm or other disaster. One of the hardest lessons learned during Hurricane Sandy, he said, was making sure that the gasoline disruptions caused by power outages do not happen again.[17]

These laws should help ensure that gasoline will be available even if the power goes out, but hurricanes can impede every stage of the fuel supply chain, from oil pumping to pipelines to refineries and retail gas stations. Stations could have electricity but still run out of gas.

Fuel availability briefly became a problem in Florida in 2005 because of a power outage at Fort Lauderdale's Port Everglades, the entry point for South Florida's fuel supply. Once that was resolved, tanker trucks were able to fill up and deliver gasoline to stations. To avoid this situation in the future, the 2006 Florida law requires motor fuel terminals and wholesalers that supply fuel to retailers to be able to operate their fuel distribution systems on an alternative generated power source for at least seventy-two hours. The alternative power source should be ready to operate within thirty-six hours of a major disaster.[18]

In September 2017, Texas motorists faced a more serious fuel supply disruption after Hurricane Harvey ravaged the region's energy distribution network, disrupting refineries and pipeline routes as well as power. Many fuel stations flooded, and stations that remained dry ran out of gasoline. Lines formed around gas stations in Austin, Dallas, San Antonio, and other Texan cities, causing dozens of locations to shut down temporarily as they awaited new supplies. Along the Gulf Coast, ten refineries in Louisiana and Texas remained closed, creating a bottleneck to process available stockpiles of crude oil. Energy providers, meanwhile, took extraordinary measures to reroute supplies to the region after the storm knocked out nearly 30 percent of the nation's refining capacity, according to the U.S. Department of Energy.[19]

Recovery turned out to be easier for EV owners after Hurricane Sandy in November 2012. In the aftermath of that storm, the *New York Times* found that many EV drivers in New Jersey were able to find charging stations in unaffected areas where power supplies had not been knocked out by the storm or were restored more quickly. In contrast, residents with gasoline cars often could not find stations with fuel, because the gasoline distribution system needs electricity not only at the retail station but also at wholesale terminals where fuel delivery trucks are filled.[20]

Micro-grids could be an ideal platform for charging EVs during power outages.[21] This approach was used successfully during Sandy, when micro-grids provided power to important sites: New York University's micro-grid was able to disconnect from the main grid and provide reliable electricity to the campus. Princeton University's micro-grid powered its campus for three days when electricity was unavailable.

In an online forum of EV drivers, many considered a home solar photovoltaic rooftop array with a storage battery pack to be the best insurance against power failures, allowing EV charging at home. They agreed that keeping a car's energy storage full is a good idea, whether it is a gasoline tank or a battery pack. A full tank of gas could get a person out of the path of the storm, and with higher-range EVs, so could a full charge.

ADAPTATION STRATEGIES

Adaptation strategies currently focus on managing and hardening the existing system to withstand climatic events. Many measures have been identified to help improve resilience of the existing energy system and avoid damages. These include data collection and analysis and use of new energy technologies to adapt to new weather risks.[22] According to the Fourth National Climate Assessment, specific actions include building and strengthening levees and floodwalls; burying power lines underground; increasing renewable energy that requires less water; deploying energy storage and micro-grid infrastructure, including demand response and islanding capabilities; improving storm readiness; and advanced planning, including securing emerging fuel supplies.[23]

In upgrading the energy system, two far-reaching goals should be addressed as part of new capital investments. First, climate change–integrated assessments show that it is imperative to transition to a new near-zero-carbon energy system within the next few decades, to slow the acceleration of climate change and to stabilize the concentration of atmospheric carbon.

Second, this new energy system should be designed to be inherently resilient to the worsening climate changes expected to come later this century, even with strong measures to cut emissions. Under the Donald J. Trump administration, the U.S. federal government has backed away from global climate agreements. But climate policy and action continue to be implemented at the subnational level, led by city and state governments, regional coalitions, organizations such as the Carbon Neutral Cities Alliance, and some industries. Goals for decarbonization and sustainability in various cities range from 80 to 100 percent decarbonization by 2050.[24]

Because energy infrastructure is long lived, decisions made now could have consequences lasting many decades. These choices need to be made despite uncertainty about technology, policy, and local effects of climate change. Questions abound about how to build a clean energy system over the next few decades and whether those systems will be more or less resilient in extreme weather events.

Clean, resilient energy systems could take many forms, depending on regional conditions and resources. For example, New York City has developed ambitious plans that include clean energy, climate resilience, and equity.[25] In the years following Hurricane Sandy, New York City's local utility Con Edison has focused on incorporating resilience as well as protective or hardening measures and predictive capabilities. New York City plans to cut greenhouse gas emissions by 80 percent by 2050. Its decarbonization efforts are focused on energy efficiency in buildings, a low-carbon grid, and more efficient transport (transport plays a smaller role than building efficiency). New York is looking at the potential for buildings to serve as virtual power plants that aggregate rooftop solar, electric cars, and building-level battery storage in ways that can deploy excess capacity to the wider grid and be more resilient to severe weather events and quicker to restore.

Decentralized energy production could provide more resilience in extreme weather events when a centralized power plant or transmission line is damaged. Geographically diverse renewable energy means the level of interruption could prove more limited, but solar panels and battery storage infrastructure could require more frequent replacement than traditional thermal plants, increasing costs. Ultimately, no single option will work for the entire United States.

Nuclear power remains a major carbon-free energy source in the United States. According to the International Atomic Energy Agency, "nuclear power plants produce virtually no greenhouse gas emissions or air pollutants during their operation and only low emissions over their full life cycle."[26] But nuclear power faces its own climate risk challenges. Potential consequences include reduced efficiency due to rising air and water temperatures, as well as power plant curtailment or shutdown under conditions of high heat and water scarcity. An increase of one degree Celsius in monthly average ambient temperature due to climate change could reduce the power capacity of the European nuclear fleet by 0.7 to 2.3 percent.[27] Like other energy facilities, nuclear power plants are vulnerable to damage during extreme weather events such as severe storms, wildfires, and floods.[28] In the wake of the Fukushima disaster, efforts were taken in the United States to harden nuclear plants to withstand catastrophic natural events.[29]

CONCLUSION

The average annual number of billion-dollar events in the United States has increased fourfold since the 1980s, and damage costs have grown by a factor of almost three and are projected to worsen.[30] At the same time, the United States has a pressing need to increase its infrastructure spending to improve water, energy, electricity, and transportation networks. As policymakers debate how to best spend federal dollars on infrastructure repair and upgrading, they need to consider the synergy between improving infrastructure resiliency to extreme climatic events and the transition to lower-carbon energy. Spending on adaptation and resiliency and low-carbon energy transitions can overlap via new technologies, such as smaller, distributed electricity grids and phone-hailed ride-sharing vans that Hawaii and New York are already testing. One interesting option is virtual power plants that aggregate power from rooftop solar panels, electric cars, and building-level battery storage and deploy the excess capacity to the wider grid.

Paying for climate damages could threaten funding for the transition to renewable energy and other low-carbon measures by creating competing requirements for public and private capital. Navigating and financing a clean energy transition in the face of growing climate disruption will require creative options from policymakers and businesses.[31] Innovative public-private structures will need to be considered in finding options for climate adaptation for the U.S. energy system.

U.S. CLIMATE RISK AND FINANCIAL MARKETS

Paul A. Griffin and Amy Myers Jaffe

Climate change poses serious and multiple risks and financial challenges to energy firms that could impede their continued access to U.S. capital markets. Climate-related physical disruptions or damages to facilities and operations from extreme weather can significantly affect future net corporate cash flows, as can rising ongoing costs for adaptation and mitigation of climate-related vulnerabilities. Another kind of climate risk faced by energy companies is transition risk. Transition risk includes risks that come from new regulatory restrictions, which could raise operating costs or inhibit the sale or use of carbon-intensive products by imposing higher-than-anticipated carbon emission penalties or taxes. In a carbon-constrained world, transition risk originates from new sources of energy and energy-efficiency equipment that could become cheaper and thereby strand carbon-intensive assets as they become nonperforming or obsolete. They also emerge when countries and large states such as California and New York transition to low-carbon-emissions economies.[1] These transition risks could harm firm value over time or all at once. Moreover, courts have exposed some energy firms to potentially large payments from climate-related legal challenges for harm or disclosure failure.[2]

These looming financial risks related to climate change for energy firms already have hit several energy companies, including coal firms and certain utilities whose stocks and bonds have suffered large devaluations related to climate-related events or trends. Climate risks can also affect energy firms' stock performance and access to credit markets when physical disruptions or severe damage to facilities and operations from extreme weather create legal liabilities or cash-flow problems that could be hard or even impossible for a company to overcome. And they could lead to market failure if investors begin to withdraw funding for a larger number of energy stocks prematurely. One perilous consequence: real-time energy shortages could result from any inability of energy firms to finance ongoing investment needed to maintain adequate and secure electrical grids and fuel supplies. Sudden drops in stock and bond prices of even a few U.S. energy businesses could create severe financial problems at these energy firms, forcing them to interrupt their ongoing sales of energy to consumers in particular locations. For example, electricity supplies have already been interrupted in Northern California after its main utility PG&E Corporation declared bankruptcy in light of potential liabilities created by its inattention to rising risks of climate-related wildfires from faulty equipment in its operating locations.[3] More such energy security problems could emerge if unexpected changes in energy company valuations—which are based on expectations of the businesses' cash flows and market worth of assets—take place rapidly, instead of over decades as markets currently assume.

At present, investors underestimate the physical and transition burdens that climate change creates for firms. This bias, which stems from many factors, could lead to substantial investor losses and spread to other parts of the financial sector, such as markets for insurance, debt, and energy commodities. Elected officials and regulators need to study this problem more carefully and consider what additional regulatory oversight is needed. The Bank of Canada, Bank of England, European Central Bank, Norwegian Government Pension Fund Global, and, most recently, a commissioner from the U.S. Commodity Futures Trading Commission have all warned of potential financial systemic risks.[4] Better policies regarding the evaluation and disclosure of corporate climate risk for energy firms are needed.

Climate change can impair corporate valuations in two ways: through physical risk, when a company faces high costs for adaptation, innovation, and mitigation in response to future climate-related events, and through regulatory and policy uncertainty, also known as energy transition risk.

Rising temperatures and sea levels caused by human-made emissions are increasing the probability of physical risks—that is, events affecting companies' assets and operations.[5] These physical factors linked to climate change can influence the amount, uncertainty, and timing of companies' future net cash flows. Investors are underestimating these harms and the risk of large potential losses for holders of those stocks and bonds. The location of energy firms and the nature of their operations can influence the extent to which investors consider climate change as a risk factor. The amount of publicly available information about the firm and the local, social, and political norms affecting managerial decision-making can also influence whether climate change risk is properly reflected in stock prices. In addition, energy firms' operations are not as geographically diversified against physical climate risks as the lofty stock valuations and favorable credit risk ratings imply. In the United States, a concentration of certain energy facilities near vulnerable coastlines or in wildfire-prone regions raises the possibility that sudden climate-related events can cause rapid, unexpected changes in valuations, as seen with the bankruptcy of PG&E.

The effect of extreme heat episodes on stock returns has been consistent with the underpricing of physical risks.[6] National Oceanic and Atmospheric Administration (NOAA) data on the frequency, duration, and cost of extreme heat episodes suggests a reduction in stock returns and an increase in stock return fluctuations.[7] Investors fail to rationally anticipate the increased likelihood and severity of future extreme weather events. Financial analysts also incorporate only part of the shock of extreme weather events into their earnings forecasts.[8] These studies support the possibility that physical climate risk is underpriced in the stock market.

The second kind of risk is energy transition risk. It stems from the uncertainty surrounding agreements, rules, and regulations that address transitioning to a low- or net-zero-carbon economy, such as the 2015 UN Climate Change Conference, also known as the twenty-first Conference of the Parties (COP21); national policies such as carbon taxes; and local regulations such as the California Global Warming Solutions Act of 2006. There is also a retrospective regulatory risk, including penalties payable to the U.S. Securities and Exchange Commission (SEC) for incomplete disclosure or court-ordered damages, such as compensation to litigants.

The stock prices of large U.S. energy companies may not reflect the full risk of carbon-intensive assets becoming stranded as a result of a transition to cleaner energy.[9] For now, investors and credit analysts generally speaking do not think the stranding of carbon-intensive energy assets is relevant to their calculations because they believe this possibility of underperformance or obsolescence will come after the shorter time frames typically under analysis.[10] However, this sanguine view ignores the risk of sudden changes in valuations based on the passage of new state or federal legislation that could alter the cash-flow expectations of the businesses. It also ignores the fact that unanticipated court decisions could similarly create sudden legal liabilities for corporations. To date, plaintiffs have been using various legal strategies to claim harm from firms' manufacturing and marketing of products or services known to contribute to rising temperatures and sea levels. Litigants also claim harm based on other avoidable damages or adaptation costs from the consequences of climate change, such as extreme weather or flooding.[11] Some fossil fuel companies are being investigated for securities fraud under the Martin Act for allegedly mischaracterizing the results of internal research and knowledge on climate change to purposely obscure to investors and the public risks to firm cash flows.[12] These legal cases add to the climate-related financial risks facing firms.

Evidence that investors and credit analysts have failed to recognize how passage of new regulations or unanticipated court cases can harm stock and bond valuations comes from financial losses that investors in coal firms have already experienced. Shifts back and forth in anticipated regulations affecting the U.S. coal industry, combined with rising competition from cheap, abundant lower-carbon energy sources such as natural gas and renewable energy, have contributed to the rapid loss of value for coal firms. The market capitalization of U.S. coal firms has already fallen precipitously by over 80 percent over the last eight years, creating large unanticipated losses for investors.[13] In addition to coal firms losing access to capital through the loss in stock valuation of their firms, over twenty banks have announced that they will no longer finance coal power firms or new coal mine projects, under pressure from climate activists. Yet coal still represents

27 percent of the energy sources for electricity production in the United States.[14] In energy security terms, it could prove problematic if investors shun coal completely even though many consumers still need it. Multiple U.S. coal firms have declared bankruptcy in the last few years.[15] This rapid decapitalization of U.S. coal firms runs the risk that such firms will not be able to meet the demand for their products in more immediate time frames when electricity from coal plants is still being used and before new generation facilities using lower-carbon energy sources can be built. In the long run, lingering advanced economies' requirements for coal could be shifted to state-financed foreign mines with higher geopolitical risk of cutoff.

One option for investors to more accurately price transition risk is to view energy transition risk as an exercise in valuing the firm's future atmospheric CO_2 emissions as an off-balance-sheet liability (a financial obligation that is not recorded on the company's official financial ledger).[16] For such a quantitative measure of transition risk, proper analysis should appropriately discount the additional financial ramifications that could ensue for each ton of reduced CO_2 emissions at all future points in time and across all possible outcomes. Priced correctly, this liability would decrease if an individual firm takes concrete actions to lower the greenhouse gas emissions linked to its operations and sales products and thereby reduce its exposure to higher future carbon taxes or restrictions. But firms that do not take actions to reduce their emissions would see a higher risk from tighter carbon restrictions in the future in the case where governments are forced to issue greater restrictions and penalties on carbon emissions down the road as the consequences of climate change worsen.

The optimal carbon price for a smooth, low-cost transition, however, is under intense debate. Absent scientific consensus, investors are uncertain how to measure the costs to firms of future required reductions in emissions. Technological changes or discoveries that could lower the costs of carbon reduction over time are also hard to assess. To date, empirical estimates of the implied average cost of carbon found on the stock market today as an off-balance-sheet liability are lower than some scientific estimates for the level of carbon pricing needed.[17] Scientists argue that the longer it takes to price CO_2 accurately, the more costly and politically difficult it could be to do so.[18] More carbon will have accumulated in the atmosphere, requiring larger, faster, and costlier reductions in emissions in the future.

Climate change regulatory uncertainty can contribute to systemic risk to U.S. financial markets as a whole.[19] Bank and investment fund losses from falling stock or bond prices in climate policy–relevant sectors, including insurers of energy companies and utilities, could destabilize the entire financial system if they happen rapidly. Regulators also should worry that climate-related financial losses, including court-ordered penalties or liabilities, could cause defaults in corporate and municipal debt markets through uncertainty in counterparty obligations—that is, the risk that one of the parties involved in a transaction could be forced to default if a change in market conditions leaves it unable to pay back a loan or another financial agreement. The same counterparty risk applies to the minimum capital requirements of the U.S. Federal Reserve for banks and other financial institutions as lenders if affected entities, such as energy companies and municipalities, cannot meet their loan obligations. For the world's largest investment fund, BlackRock, this kind of counterparty risk exposure approximates $1.4 trillion.[20]

The sobering case of PG&E is emblematic of these various kinds of climate risks that can bring unexpected losses for investors and interrupt energy supplies. First, PG&E's apparent inattention to the physical risks it faced from rising climate-related risk of wildfires in its operating locations caused investors in its stocks and bonds to suffer large losses. California's courts could possibly rule that PG&E is strictly liable for tens of billions of dollars in wildfire damages under inverse condemnation, a legal principle that the company can be held fully responsible because its operations destroyed life and property while performing its public function.[21] These potential legal liabilities and the firm's related subsequent bankruptcy declaration have virtually eliminated the company's access to private capital markets. This has created energy security problems for Northern California. Because of PG&E's financial problems, it is denying energy to certain service areas temporarily when fire risk is high. PG&E announced it will turn off electricity for customers when heat-wave and wind conditions raise the risk that an equipment failure could cause a fire,[22] as an interim step to address its lack of access to financing to make necessary repairs and upgrades to its equipment. PG&E's bankruptcy also threatens future energy supply for Northern California because the firm likely cannot pay for existing contracts for ongoing renewable energy projects in the state. The state of California was also counting on PG&E to build electric car charging stations and to achieve 100 percent renewable energy generation for its electric grid.[23]

Second, in an example of how PG&E's inattention to climate risk is going beyond its balance sheet issues, its massive liabilities threaten the continued functioning of California's insurance market.[24] In July 2018, the California state legislature passed a bill establishing a $20 billion fund to pay for wildfire liabilities faced by the state's utilities. But it is not clear if the plan, which includes retail rate increases aimed to bring PG&E out of bankruptcy, will be sufficient to address both the existing liabilities and the future risks. PG&E has stated that the inspections, repairs, and upgrades needed to make its infrastructure and operations safe to California residents could cost as much as $75 billion to $150 billion.[25]

EXPLANATIONS OF CLIMATE RISK UNDERPRICING

Several sources provide information on how climate change could affect company cash flows. But constraints and frictions on information production, inadequate disclosure rules, and investor inattention provide some explanation for why climate risk remains under-assessed. One problem is that investors may not have access to fully accurate information. There is mandatory and voluntary reporting, including disclosures on how companies intend to comply with COP21 guidelines. As recent litigation has exposed, internal company documents can be another source of information.[26] Although these kinds of climate change information mean climate risk information is not in short supply, the information companies provide to investors, creditors, and customers could still be insufficient to prevent biased or faulty analysis that downplays possible risks. For example, some analysts argue that corporate disclosures are improperly downplaying the high risk that oil reserves could drop in value if governments severely restrict fossil fuel use, often referred to as stranded-asset risk.[27] Indeed, studies indicate stranded-asset risk could be not yet fully priced into fossil-fuel-firm market values.[28]

Firms maintain that plans to address transition by selling assets the firms think could lose value over time are too sensitive for public disclosure because revealing this information in advance could harm the eventual value of the asset or enterprise. This failure to disclose internal knowledge regarding problematic assets means energy company stock prices could reflect overestimation of firm value. Some lower courts have upheld energy firms' right to restrict disclosure of confidential information, and the SEC can only subpoena internal documents as part of a formal enforcement action. But legal precedents could change. Recently, the U.S. Supreme Court ruled in favor of discovery by plaintiff Maura Healey of documents dating back to

January 1, 1976, relating to ExxonMobil's possible knowledge about climate change and global warming.[29]

Climate science studies contribute to closing some of these informational deficits, but that kind of information is harder for investors to apply to specific company holdings and activities. Prices of contracts for transferring climate risk to others through derivatives and insurance markets also provide some aggregated information.[30] Some investors rely on climate disclosure advocacy groups, such as CDP Worldwide and Ceres, to glean additional information. They also tap reports from analysts and research organizations, including MSCI ratings, credit rating agencies, and proprietary datasets, but the persistence of underpricing of risk indicates that different kinds of data are needed. These nongovernmental organizations' actions to promote additional firm-level data have produced somewhat more information, including two-degree-scenario analyses by some fossil fuel firms. But most firms limit their climate risk disclosures to boilerplate language, which is insufficient for regulators, policymakers, and investors to adequately assess potential financial risks.

Climate risk prediction markets could provide better aggregated information for investors and add additional transparency to how market participants price the probability of scientific projections, such as future sea-level rise or heat-wave incidence. Such tradable risk instruments, moreover, could be useful for hedging and insurance purposes. These markets could add transparency by allowing participants to value accurately the best available scientific data on climate change effects, such as rising sea levels and temperatures.[31]

Climate risk underpricing could also result from investor inattention. This phenomenon can increase the propensity for herding or an information cascade, wherein market participants follow an observable trend instead of seeking more accurate information or trusting their own knowledge.[32] When some investors have private knowledge of the accurate degree of climate risk but fail to trade on that information (that is, reflect it in market prices), a cascade can be created and later triggered by an unblocking event. For example, investors could fail to fully price the effects of stranded-asset risk—a clear and significant risk, according to climate scientists and policy experts—until the occurrence of an event that they can no longer ignore, such as a stricter, credible global agreement committing all countries to limit fossil fuel production. Investor inattention could also be the result of the sheer volume of climate risk information, because information overload can lead analysts to delay including that information in earnings forecasts.[33]

When assessing energy transition–related risks, investors' existing underpricing of risks could be based on the idea that regulation is currently weak. This could prove to be faulty logic if more stringent regulations with greater effects on company earnings arise in the future. This timing factor makes it trickier for investors to price or hedge the risk. One example is the green paradox: the risk that a high likelihood of a future tax or restriction on carbon will encourage firms to protect profits by accelerating production ahead of the tax and thereby increase company emissions in the short run.[34]

Another reason why investors tend not to worry: they assume a long-tail risk such as climate change means businesses' prospective cash-flow declines would occur in the distant future, whether those effects on cash flow involve physical damage to assets from weather and sea-level rise or a loss of future market share for high-carbon products. This implies a trade-off between the high value of the firm's assets, products, and services from today's cash flows and a potentially much lower value in the distant future. When managers have shorter-run incentives, they will discount long-run cash flows at a much higher rate. In effect, they ignore the harm of distant future outcomes, leaving future risks beyond twenty years largely unpriced or unmanaged.[35] Economists have debated a long-run discount rate for future climate assessment fervently but without much agreement. Such rates are subject to widely varying assumptions about the trade-offs of economic activity that produces higher cumulative carbon emissions now versus higher reductions of greenhouse gases in the future.[36] However, as the case of PG&E and U.S. coal firms has already shown, investor losses can be more immediate than commonly assumed.

CONCLUSION

Investors are underestimating climate risk from both the physical and transition burdens that it could create for firms. This bias, stemming from many complex factors, is creating financial risks that could affect reliability of energy supply, lead to substantial investor losses, and spread systemic failure to other parts of the financial sector such as markets for insurance, debt, and energy commodities. Sudden sharp drops in stock and bond prices of a small handful of U.S. energy businesses over the last few years have curtailed the energy supply for consumers of specific geographic locations, and created substantial investor losses. Some investment firms are starting to recognize that prices have not been fully discounted for climate risk. In some cases, they are building portfolios of stocks with low-carbon risk that they hope will yield returns superior to a comparative broader index such as the S&P 500.[37] But government regulators are not doing enough to mitigate the possibility that risks could cascade through U.S. financial and energy commodity markets.

Finding the right level of disclosure to reduce the risks of biased and faulty information has so far proven elusive. Detailed information on corporate strategies and investment plans is largely unavailable, in part because disclosing such information can create competitive disadvantages and increase litigation exposure. Companies fear that increased recognition of climate risk as a valuation factor will prompt onerous regulation, so disclosure statements tend to stick with legal boilerplate that is vague and uninstructive. Additional regulation is necessary given the harm that can come from energy supply outages caused by sudden losses in financial solvency of energy firms. The SEC should revisit its almost ten-year-old guidance statement on the topic to address new concerns.[38] Congress, the U.S. Commodity Futures and Exchange Commission, and state governments should also consider which individual company disclosures are needed to assess whether climate risk in financial markets represents a major threat to localized or national energy supply, the proper functioning of energy commodity and insurance markets, and the stability of financial markets.

A CLEAR AND PRESENT DANGER
Climate Risks, the Energy System, and U.S. National Security

Joshua Busby

In August 2017, Hurricane Harvey dumped more than forty inches of rain on the Houston area in four days, as high as sixty inches in some places.[1] Three hundred thousand Texans lost power, and the effects cascaded to critical infrastructure, including hospitals, wastewater treatment plants, and refineries. Like Hurricanes Katrina and Rita in 2005, Harvey shut down refineries and oil production in the Gulf of Mexico, which account for some 11 percent of total U.S. refining capacity and a quarter of oil production. In Harvey's wake, fears of shortages led to a spike in gas prices through the South and the country.[2] The country barely had time to come to grips with the effects of Harvey when Hurricane Irma buffeted Florida and knocked out two-thirds of the power supply to Floridians in early September.[3] For both storms, the Coast Guard and the National Guard, along with other military assets, were activated in the tens of thousands to provide emergency relief.[4]

Days later, a third storm would stretch and test an overburdened disaster-response system further. In mid-September, Hurricane Maria destroyed the power grid on the island of Puerto Rico, leaving much of the U.S. territory in darkness for months on end. It took eleven months for power to be fully restored to 1.5 million customers, making the blackout the lengthiest in American history.[5] One of the tasks U.S. civilian and military responders faced in the immediate wake of the storm was to try to restore power, particularly to critical locations such as ports, airports, and hospitals.[6] Intermittent and incomplete power service were part of the reason nearly three thousand Puerto Rican Americans died in the wake of the storm, faced with increased exposure to searing heat and rainfall as well as the lack of power for medical equipment.[7] The island experienced more crime in the wake of the storm, including an increase in murders and theft of generators, with police finding it more difficult to patrol in areas experiencing power outages.[8]

A 2013 study for the U.S. Department of Energy warned of a variety of climate effects on the energy system, including thermal power plant vulnerability to droughts and high temperatures; vulnerability of coastal infrastructure, including cities, refineries, electricity grids, and pipelines, to storms; oil and gas vulnerability to declining water availability; effects of changing water availability on renewable energy systems, particularly hydropower; risks to energy transmission and distribution from high temperatures and wildfires; disrupted barge and rail traffic due to drought and flooding; and changes to the Arctic, including the effects of melting permafrost on infrastructure.[9] Other effects, such as power outages, were observed in 2018 during the polar vortex, when spikes in demand led to gas-plant fuel shortages in some parts of the country and ice-clogged cool water intakes at nuclear power plants in New Jersey and Delaware, forcing them offline.[10]

These outages and effects cause economic damage and disrupt the lives of Americans, but the consequences can also rise to the level of a U.S. national security concern. Climate consequences for the energy sector could constitute national security threats through a variety of pathways. Policymakers should take these concerns seriously and prepare for these risks.

CLIMATE CHANGE AND NATIONAL SECURITY

Distinctions are often made between direct threats to the homeland and indirect threats to a country's national interests coming from climate change.[11] Here, the focus is on direct threats to the homeland. The limited classic study of national security involves external attacks by armed foreign adversaries, namely nation-states. That does not leave room for nonstate actors, nor does it create space for threats that could rise to the level of security consequences but not be carried out on purpose by human agents.

It is tempting to see any harms to the military as national security threats, which narrows the risks to the most important institution charged to protect countries from threats. The U.S. Department of Defense (DOD) has carried out a number of studies to assess the vulnerability of bases to climate harms and the ways in which climate change could shape military missions, operations, and training. A specific inventory now exists of the extent of risks to military installations; nearly half of some 3,500 installations, according to a 2018 congressionally mandated report, faced some sort of climate hazard exposure.[12] An accounting of particular risks to seventy-nine critical military installations was undertaken as part of a recent 2019 congressionally mandated study.[13] Some bases, such as those in Norfolk, Virginia, face flood risks. Given sea-level rise, they could have to make multibillion-dollar investments to remain viable. Others are subject to extreme weather events and wind, as Tyndall Air Force base experienced during Hurricane Michael. Still others face risks of drought and high temperatures that pose fire risks, which could preclude live ammunition training. Other installations, notably those in Alaska, face risks from melting permafrost and coastal erosion that are destabilizing structures, including major radar stations.

However, these installations depend on the wider civilian communities in which they are embedded, particularly because they rely on civilian electricity grids for power. Analyses are needed that expand on the threats to civilian energy systems and what they mean for the U.S. military. These extended threats to U.S. military installations at home and abroad constitute national security threats by virtue of their importance to force preparedness and deployment, but they are not the full extent of threats.

Focusing on what climate change means for just the U.S. military excludes severe harms to the nation, such as pandemic disease, that in the modern era are considered national security threats. In considering what makes pandemic diseases and climate change security threats, the first factor is the severity of harms to the country. A second dimension is the speed with which they potentially occur, giving the country too little time to minimize harmful consequences. National security has historically been defined as existential threats to the government, but expansive concepts, such as human security, emphasize the threats to individual well-being. The potential for large-scale loss of life from exposure to climate hazards and their aftermath constitutes a security risk in its own right. The destruction and disruption could also rise to the level of security threats if they significantly diminish or challenge the country's way of life or the way of life of a sizable proportion of the country.

A number of harmful potential results of climate risks could rise to a national security threat. First, the deaths of large numbers of citizens would constitute unnecessary and unacceptable human suffering in democratic societies, whose mission is to serve the people.

Second, extreme climate risks would expose the political leadership to legitimacy deficits that could undermine their ability to govern and cause them to lose support from important constituencies. In nondemocratic contexts, failure to respond to severe weather events can constitute threats to the regime leadership and their continued rule.[14] In the United States, the political fallout in the wake of Hurricane Katrina was encapsulated by musician Kanye West's remark that "George Bush does not care about black people."[15] President Donald J. Trump was praised for overseeing effective responses to Hurricanes Harvey and Irma, but his response to Puerto Rico in the wake of Hurricane Maria was harshly criticized.

Third, extreme weather events increasingly require military mobilization to prevent loss of life, deliver relief, reestablish order, restore electricity, and make other critical infrastructure operational. The opportunity costs that divert military assets, decision-makers, and national security personnel from their other core missions thus constitute security trade-offs. For example, a military unit's assisting on a climate-related humanitarian mission can affect readiness because the deployed troops then need to be resettled and retrained upon their return. Military mobilization for humanitarian response has been seen not only with hurricanes but also in response to the wildfires that California experienced in 2018, the floods throughout Nebraska in 2019, and other events.

Fourth, the damages from climate hazards impose unacceptable costs when they damage civilian infrastructure. Hurricanes Harvey, Irma, and Maria together were estimated to cause damages in excess of $200 billion.[16] This is on par with the physical and economic damage from the attacks of September 11, 2001.[17] Although the intention of the attackers is part of the reason terrorism is a national security threat, intentionality is not inherent to such threats. President Barack Obama committed billions of dollars and dispatched thousands of troops to West Africa to fight the Ebola epidemic in 2014.[18] In doing so, he evoked national security threats and the possible ripple effects on loss of life and the global economy: "And if the outbreak is not stopped now, hundreds of thousands of people could potentially become infected, with profound political and economic and security implications for all of us."[19] This was recognition that a disease could pose unacceptable costs on the United States and the world. By the same token, if the damage from the 2017 hurricanes had been exacted by a human agent such as al-Qaeda, the United States would have been prepared to go to war. Intentionality creates a perpetrator who can be held responsible, but unintentional harms can be comparable in severity.

Fifth, although climate change may not be an existential challenge for the entire United States, it is one for parts of the country. Small communities have become uninhabitable due to sea-level rise, and Americans already have to be moved from parts of Alaska, Louisiana, and other climate-vulnerable areas.[20]

Climate change effects have a human element that goes beyond the energy sector. For example, Hurricane Irma knocked out the electricity grid on the Caribbean island of Barbuda (not a U.S. possession), which required the emergency evacuation of all 1,700 inhabitants.[21] Other island communities, such as American Samoa, Guam, the Northern Mariana Islands, Puerto Rico (again), Saipan, and the U.S. Virgin Islands, could experience such knockout blows to the power grid and wider infrastructure. Some of these places could eventually be considered unlivable for all or a portion of their inhabitants, and not just temporarily. As larger parts of the country become uninhabitable (in part because of damage to energy systems or temperatures that render cooling systems vulnerable to persistent blackouts), the human security consequences of these movements could become larger national preoccupations.

THE PATHWAYS TO SECURITY CONSEQUENCES IN THE ENERGY SECTOR

Climate change could create national security consequences through its effects on the energy sector in many ways. In considering climate change and national security, the cascading nature of energy disruptions needs to be better understood. Traditional approaches to energy security and cyber defense acknowledge these risks.

Military Base and Wider Community Vulnerability

Military bases could find their own operations at risk from civilian energy-sector vulnerabilities during climate-related emergencies. As the 2017 Quadrennial Defense Review noted, the "Department of Defense (DOD) is the largest customer of the electric grid in the United States, a system that is largely owned and operated by the private sector."[22]

For example, Camp Lejeune experienced power outages during Hurricane Florence in 2018. Camp Lejeune had backup power from generators, but the base also had to house civilians who themselves lost power and needed refuge.[23] This shows how wider civilian vulnerabilities affect military installations. Soldiers' families are often housed in the local community, so disruptions to civilian energy generation could matter for base operations and have wider implications for the base. As the former commander of Langley Air Force Base, retired General Ron Keys, noted,

> Now I can build a moat, or a barrier around Langley Air Force Base, but the problem is a lot of my people live in Newport News, live in Hampton. A lot of my electricity comes in from outside. My fuel comes in from outside. So at some point we get to the point: "I've got to move to higher ground."[24]

The 2019 congressional study discussed vulnerabilities beyond the electricity grid. Data servers could be subject to storm or temperature-related outages, which could impair wider functionality of critical national security decision-making and situational awareness around the world. The Defense Logistics Agency reviewed cooling capacity of its servers and moved some operations from flood-prone areas to higher ground. It is unclear how exhaustively the 2018 or 2019 studies considered the wider risks of community energy-sector disruptions for bases. The DOD is aware of these risks and, aside from having backup generators, has looked for opportunities to work with the private sector on energy resilience. Given funding constraints, it is also open to new options, such as using buildings as a backup energy source via microgrids and batteries.

Humanitarian Emergencies

The military could increasingly be called upon for domestic humanitarian response in the wake of climate emergencies. The international dimensions are discussed in the 2019 study, but the domestic humanitarian piece is reduced to a paragraph because, as the report notes, the Federal Emergency Management Agency (FEMA) and other civilian agencies lead these operations and all anticipatory hazard planning. FEMA's 2018 four-year strategic plan removed language about climate change. Although the U.S. military continues to think about and prepare for the threats of climate change in a limited way, wider security consequences are being ignored or dealt with obliquely without explicitly mentioning climate change.[25] This security threat has two dimensions: the risks of civilian death from exposure to climate hazards and the need for military mobilization to prevent such losses.

Domestic Cascading Effects

The existing interrelationship of critical infrastructure systems could lead to far greater damages to quality of life and the U.S. economy, as damage in one area can have cascading effects on other sectors or regions. These risks were documented in the Fourth National Climate Assessment, which notes that "systems that depend on one another are subject to new and often complex behaviors that do not emerge when these systems are considered in isolation. These behaviors, in turn, raise the prospect of unanticipated, and potentially catastrophic, risks."[26] Cascading effects from climate change that have already taken place include health challenges caused by flood waters and toxic releases from Hurricane Harvey, looting during blackouts in New York City, and erosion and permafrost thaw in Alaska, which has led to collapse of bridges, buildings, and access to transportation routes. Tightly connected supply chains for vital goods can also transmit problems from one region of the world to others.

In the wake of September 11, Thomas Homer-Dixon, a scholar of environmental security, warned of these consequences in his assessment of terrorism and the concentration of value in urban areas where human populations, economic and financial activity, electricity generation, information networks, and transport hubs are all located.[27] Although interdependence has its virtues, such as just-in-time delivery, tightly coupled networks are subject to cascading disruptions in the same geographic domain and beyond.

A 2016 study from the U.S. Department of Energy noted that Hurricane Sandy affected not only the electricity sector but also the transport sector: the New Jersey Transit operations center that controlled a variety of critical functions flooded.[28] The 2017 Quadrennial Energy

Review noted that Sandy also knocked out communications systems that depend on electricity.[29]

Cascading effects occur not only geographically and functionally, across regions and sectors (electricity to heating to hospitals, ports, airports, sanitation, and so on), but also temporally. In 2017, three major storms in succession overwhelmed FEMA's capacity to respond effectively, which depleted resources and the capacity of the organization to guide multiple ongoing emergencies at the same time. The U.S. military has a doctrine of being able to fight two major wars simultaneously. Analogously, FEMA needs enhanced capacity to be able to address overlapping severe emergencies. The 2018 FEMA strategic plan, despite its flaws, recognized the need for surge capacity and greater coordination with other agencies.

The identification of cascading risks should go beyond swift-onset disasters to more slow-moving processes, such as temperature, rainfall change, and changes in seasonality, that could make it impossible to heat, cool, light, or move energy resources throughout a given area, whether because of too little water in major hydroelectric dams, melting permafrost that leads to damage to pipelines, persistent fire risks, sea-level rise, or perennial flood risks that make electricity assets unusable.

International Cascading Effects

There could be additional cascading risks when U.S. energy markets are bound up with other countries—for example, using energy resources that come from hydropower or fossil fuels from Canada or other parts of the world. The U.S. transportation sector is moving or could move toward electric vehicle scale-up backed by renewables, but imported products such as solar panels and critical minerals could be subject to supply disruptions if they are sourced from climate-vulnerable areas in other parts of the world. Such disruptions have already affected global supply chains and caused billions of dollars in losses.[30] Certainly, the risks to U.S. exports from vulnerable petrochemical complexes located along the Gulf Coast are large and could have ripple effects on global markets.

CATEGORIZING HAZARD EVENTS

Differentiating between a run-of-the-mill extreme weather event and a national security extreme weather event is challenging, given that even smaller hazards often require some measure of military mobilization through the National Guard as part of an effective response. However, a checklist of questions can be used to distinguish lower-level national security risks from more extreme ones (see figure 1). The more yeses to these questions, the higher the degree of concern that a single hazard event or several in close succession would pose higher-level national security risks. Thus, the most severe national security–climate–energy system risk would be a series of climate hazard events that simultaneously threatened the energy systems of bases, communities near bases, critical bases, and population centers; required military mobilization; posed cascading risks for wider infrastructure; overextended the disaster response system; and were interlinked with wider, overlapping climate hazards to international energy sources, markets, and grids.

ADDRESSING NATIONAL SECURITY RISKS

Policymakers have a number of options to address these risks.

First, Congress should direct the military to assess the full scope of energy-sector risks from climate change to civilian communities where it operates. (It is possible that it has, in the classified and extended versions of reports such as the DOD 2019 report, but this would be a start if such an assessment has not been carried out.[31]) This assessment should cover at least the seventy-nine core facilities in the 2019 report.

Second, Congress should direct FEMA and other agencies to carry out or update a similar risk map for urban centers, including but not limited to energy-sector risks for extreme weather events and climate change.[32] Congress should task FEMA and the Department of Defense with responding to that risk map with an assessment of what the findings could mean for the military. This could involve scenario development to anticipate what climate disasters could mean for military mobilization. The map should identify the locations most vulnerable to climate-related humanitarian emergencies, with some breakdown by hazard type. It should also include the extent of vulnerability in terms of population and potential damages, with priority locations identified in terms of the scope for large-scale loss of life, large-scale economic costs, and potential for cascading consequences on financial markets or energy prices in the rest of the country. These reports should be a move toward a list of action items and cost estimates to reduce local vulnerabilities through measures such as hardening investments, backup micro-grids, battery storage, and other efforts. The work should cover risks that could happen over the next decade or two, as longer time-horizons offered by climate models are less useful for present-day planning. At the same

Figure 1. NATIONAL SECURITY, CLIMATE, AND ENERGY SECTOR HAZARD EVENT RANKING

Question	Points (out of 8)
Military bases and communities	
1. Does a climate hazard pose a threat to military bases' energy systems?	
2. Does a hazard impose wider risks to bases via their effects on energy systems in communities where they are colocated?	
3. Are those bases operationally central to U.S. national security?	
Domestic humanitarian emergency	
4. Do climate risks threaten the energy systems of large population centers?	
5. Would military assets be required for humanitarian rescue and service restoration?	
Domestic cascading effects	
6. Could a hazard have cascading effects on the energy system and other systems and generate unacceptable economic losses?	
7. Could overlapping hazards in succession impose unacceptable strain on disaster response capabilities?	
International cascading effects	
8. Could a hazard event in another country generate cascading effects on the U.S. economy?	

Source: Joshua Busby.

time, such an assessment ought to combine historic hazard exposure with emergent understandings of climate change phenomena such as rainfall stalls over land and rapid intensification of storms.[33]

Third, the Department of Energy, Department of Homeland Security, and other agencies should revisit Obama-era analyses on climate risks to the energy sector. The electricity sector has experienced dramatic change in recent years with the penetration of natural gas and renewables and, to a lesser extent, the decline in coal and nuclear plants. The contemporary cascading climate risks to the energy system, including electric power generation, should be assessed with a list of priority action items and cost estimates.

Fourth, the Department of Energy should assess the degree to which the U.S. energy sector could be subject to power or fuel disruptions resulting from climate change, either as a result of international disruptions to imported energy or energy-related products (such as solar panels) or as a consequence of climate disruptions to U.S. energy exports.

CONCLUSION

Energy-sector risks from climate change for bases (and surrounding communities) are the most obvious starting points for action, building off the 2018 and 2019 studies. A more challenging assessment would identify the metropolitan areas most at risk from climate-related humanitarian emergencies and the resource and organizational implications for different parts of the U.S. government, including the military. A further step would require assessing the extent to which international climate disruptions could have an effect on U.S. energy markets domestically or the extent to which disruptions to U.S. energy markets could have ripple effects internationally.

Together, such analytical work could set the stage for productive priority setting and an inventory of actionable investments to shore up U.S. climate resilience.

ENDNOTES

CLIMATE CHANGE, STORM SURGE, AND THE OIL AND GAS INDUSTRY

1. Benjamin Strauss, Remik Ziemlinski, Jeremy Weiss, and Jonathan Overpeck, "Tidally Adjusted Estimates of Topographic Vulnerability to Sea Level Rise and Flooding for the Contiguous United States," *Environmental Research Letters* 7, no. 1 (March 2012), http://researchgate.net/publication/231072679_Tidally_adjusted_estimates_of_topographic _vulnerability_to_sea_level_rise_and_flooding_for_the_contiguous_United_States.

2. Robert Kopp, Radley Horton, Christopher Little, Jerry Mitrovica, Michael Oppenheimer, D. J. Rasmussen, Benjamin Strauss, and Claudia Tebaldi, "Probabilistic 21st and 22nd Century Sea-Level Projections at a Global Network of Tide-Gauge Sites," *Earth's Future* 2, no. 8 (June 2014), http://agupubs.onlinelibrary.wiley.com/doi /full/10.1002/2014EF000239.

3. Anthony Westerling, "Wildfire Simulations for California's Fourth Climate Change Assessment: Projecting Changes in Extreme Wildfire Events With a Warming Climate," State of California Energy Commission (August 2018), http://climateassessment.ca.gov/techreports/docs/20180827-Projections_CCCA4-CEC -2018-014.pdf.

4. "Hurricanes Maria, Irma and Harvey Situation Reports: Archived: August 26, 2017–September 19, 2017," U.S. Department of Energy (DOE) (September 2017), http://energy.gov/ceser/downloads/hurricanes-maria-irma-and-harvey-situation-reports -archived-august-26-2017-september.

5. Krista L. Jankowski, Torbjorn E. Tornqvist, and Anjai Fernandes, "Vulnerability of Louisiana's Coastal Wetlands to Present-Day Rates of Relative Sea Level Rise," *Nature Communications* 8, (March 2017), http://dx.doi.org/10.1038%2Fncomms14792.

6. John Radke, Greg Biging, and Karlene Roberts, "Assessing Extreme Weather: Related Vulnerability and Identifying Resilience Options for California's Interdependent Transportation Fuel Sector," California State Energy Commission (August 2019), http://climateassessment.ca.gov/techreports/docs/20180827-Energy_CCCA4-CEC -2018-012.pdf.

7. Frank Bajak and Lise Olsen, "Silent Spills: Part 1: In Houston and Beyond, Harvey's Spills Leave a Toxic Legacy," *Houston Chronicle*, n.d., http://houstonchronicle.com/news/houston-texas/houston/article/In-Houston-and-beyond-Harvey-s-spills-leave-a-12771237.php.

WATER-RELATED RISKS AND IMPACTS ON THE U.S. ENERGY SYSTEM

1. Intergovernmental Panel on Climate Change (IPCC), "2018: Summary for Policymakers," in *Global Warming of 1.5°C*, ed. Valérie Masson-Delmotte et al. (Geneva: IPPC, 2019), http://ipcc.ch/site/assets/uploads/sites/2/2018/07/SR15_SPM_version_stand_alone_LR.pdf.

2. Kevin Krajick, "Scientists See Fingerprint of Warming Climate on Droughts Going Back to 1900," *Lamont-Doherty Earth Observatory*, May 1, 2019, http://ldeo.columbia.edu/news-events/scientists-see-fingerprint-warming-climate-droughts-going-back-1900.

3. Debra Perrone and Scott Jasechko, "Dry Groundwater Wells in the Western United States," *Environmental Research Letters*, September 28, 2017, http://iopscience.iop.org/article/10.1088/1748-9326/aa8ac0.

4. Jason Dearen and Michael Biesecker, "Toxic Waste Sites Flooded in Houston Area," Associated Press, September 3, 2017, http://apnews.com/27796dd13b9549b0ac76aded58a15122.

5. Zachary Sadow, Daniel Ford, Mark Lewis, Rose-Lynn Armstrong, Upmanu Lall, Michelle Ho, and Angela LaSalle, "The Water Challenge: Preserving a Global Resource," *Barclays*, March 22, 2017, http://investmentbank.barclays.com/content/dam/barclaysmicrosites/ibpublic/documents/our-insights/water-report/ImpactSeries_WaterReport_Final.pdf.

6. Russell Gold, "Energy Firm Makes Costly Fracking Bet," *Wall Street Journal*, August 13, 2013, http://wsj.com/articles/energy-firm-makes-costly-fracking-beton-water-1376438982.

7. Poulomi Ganguli, Devashish Kumar, and Auroop Ganguly, "U.S. Power Production at Risk From Water Stress in a Changing Climate," *Scientific Reports* 7, article 11983 (2017), http://nature.com/articles/s41598-017-12133-9#ref-CR16.

8. "Wood MacKensie: Lack of Water Is a Risk to Global Energy Industry," *Offshore Technology Today*, November 8, 2013, http://offshoreenergytoday.com/wood-mackenzie-lack-of-water-is-a-risk-to-global-energy-industry.

9. Sadow et al., "The Water Challenge."

10. Ibid.

11. For example, the International Energy Agency says that global water consumption for power generation and fuel production is expected to more than double from 66 billion cubic meters (bcm) in 2010 to 135 bcm by 2035. Coal accounts for 50 percent of this growth and is one of the most water-intensive methods of generating electricity.

12. Individual companies scored include: Ameren Corporation, American Electric Power Company, Anadarko Petroleum Corporation, CMS Energy Corporation, CONSOL Energy Inc., Devon Energy Corporation, Dominion Energy, DTE Energy, Duke Energy Corporation, EOG Resources Inc., Exelon Corporation, Halliburton, NISource Inc., NRG Energy Inc., Occidental Petroleum Corporation, PG&E Corporation, Pinnacle West Capital Corporation, Sempra Energy, The AES Corporation, and WEC Energy Group. Details on the scores and methodology are available on the CDP website. For further information on CDP scoring of corporate responses, see CDP's Scoring Introduction 2019, https://6fefcbb86e61af1b2fc4 -c70d8ead6ced550b4d987d7c03fcdd1d.ssl.cf3.rackcdn.com/cms/guidance_docs /pdfs/000/000/233/original/Scoring-Introduction.pdf?.

13. The reported worst-case scenario was 1.4 meters of sea-level rise coupled with a hundred-year storm event.

14. Adam Freed, "Message From Hurricanes Michael and Maria: Renewable Energy Makes More Sense Than Ever," *USA Today*, October 14, 2018, http://usatoday.com/story /opinion/2018/10/14/hurricane-michael-maria-renewable-energy-infrastructure -sustainable-solar-wind-column/1575967002.

15. Irina Ivanova, "Hurricane Florence crippled electricity and coal—solar and wind were back the next day," *CBS News*, September 25, 2018, http://cbsnews.com/news/hurricane -florence-crippled-electricity-and-coal-solar-and-wind-were-back-the-next-day.

CLIMATE CHANGE IMPACTS ON CRITICAL U.S. ENERGY INFRASTRUCTURE

1. Craig Zamuda et al., "Energy Supply, Delivery, and Demand," in *Impacts, Risks, and Adaptation in the United States: Fourth National Climate Assessment*, vol. 2, ed. David R. Reidmiller et al. (Washington, DC: U.S. Global Change Research Program, 2018), doi: 10.7930/NCA4.2018.CH4, http://nca2018.globalchange.gov/chapter/energy.

2. Zamuda et al., "Energy Supply," 176; U.S. Department of Energy (DOE), *Climate Change and the U.S. Energy Sector: Vulnerabilities and Resilience Solutions*, DOE/EPSA-0005 (Washington, October 2015), 1, http://energy.gov/sites/prod/files/2015/10/f27 /Regional_Climate_Vulnerabilities_and_Resilience_Solutions_0.pdf; Adam Smith, Neal Lott, Tamara Houston, Karsten Shein, Jake Crouch, and Jesse Enloe, "U.S. Billion-Dollar Weather and Climate Disasters 1980–2019," National Oceanic and Atmospheric Administration: 1–15, http://ncdc.noaa.gov/billions/events.pdf.

3. Smith et al., "U.S. Billion-Dollar Weather."

4. Michelle Davis and Steve Clemmer, "Power Failure: How Climate Change Puts Our Electricity at Risk—and What We Can Do," Union of Concerned Scientists (April 2014): 7, http://ucsusa.org/powerfailure.

5. "Climate Change: Energy Infrastructure Risks and Adaptation Efforts," U.S. Government Accountability Office-14-74, January 31, 2014, 24, http://gao.gov/products /GAO-14-74.

6. Robert Watson, James McCarthy, and Liliana Hisas, "The Economic Case for Climate Action in the United States: FE-US, 2018," Universal Ecological Fund, 2, http://feu-us .org/case-for-climate-action-us.

7. Davide Banis, "Counting the Cost: A Year of Climate Breakdown 2018," Forbes, December 28, 2018, http://forbes.com/sites/davidebanis/2018/12/28/10-worst-climate -driven-disasters-of-2018-cost-us-85-billion/#7a4c19c22680.

8. Zamuda et al., "Energy Supply," 179.

9. Tanishaa Nadkar, Noor Zainab Hussain, and James Emmanuel, "Insured Losses From Camp and Woolsey Wildfires Estimated at $9–13 Billion: RMS," Reuters, November 19, 2018, http://reuters.com/article/us-california-wildfires-insurance/insured-losses -from-camp-and-woolsey-wildfires-estimated-at-9-13-billion-rms-idUSKCN1NO18Y.

10. Dale Kasler, "A Statewide Problem. How PG&E's Bankruptcy Could Soil California's Green-Energy Movement," *Sacramento Bee*, January 31, 2019, http://sacbee.com/news /business/article224750505.html.

11. Zamuda et al., "Energy Supply," 184.

12. DOE, *Quadrennial Energy Review: Transforming the Nation's Electricity System: The Second Installment* (Washington, 2017) S-6, http://energy.gov/epsa/quadrennial-energy -review-second-installment.

13. Daniel Sperling, *Three Revolutions: Steering Automated, Shared, and Electric Vehicles to a Better Future* (Chicago: Island Press, 2018).

14. "Energy Technology Perspectives 2017: Executive Summary," International Energy Agency, 4, http://webstore.iea.org/download/summary/237?fileName=English-ETP -2017-ES.pdf.

15. Jim Blackburn and Amy Myers Jaffe, "Climate Change, Storm Surge, and the Oil and Gas Industry," in *Impact of Climate Risk on the Energy System*, ed. Amy Myers Jaffe (New York: Council on Foreign Relations, 2019).

16. Alexandra Seltzer, "Storm-Proofing the Pumps: Most Have Switched to Generator Power," *Palm Beach Post*, July 11, 2010, http://palmbeachpost.com/news/stormproofing -the-pumps-most-have-switched-generator-power/fH2khwqA42eqN6fZihyLpI.

17. "Governor Cuomo Announces More Than 250 Downstate Gas Stations Installing Back-Up Power Capacity to Prepare for Future Emergencies," Office of the Governor, State of New York, October 29, 2013, http://governor.ny.gov/news/governor-cuomo-announces -more-250-downstate-gas-stations-installing-back-power-capacity-prepare.

18. Seltzer, "Storm-Proofing the Pumps."

19. Laura Santhanam, "Vehicles Rush Texas Gas Stations as Supplies Dwindle After Harvey," PBS *Newshour*, September 1, 2017, http://pbs.org/newshour/nation/vehicles -rush-texas-gas-stations-supplies-dwindle-harvey; Nathan Bomey, "Gas Shortages Crop Up in Texas After Hurricane Harvey Stirs Fears," *USA TODAY*, September 1, 2017, http://usatoday.com/story/money/2017/09/01/hurricane-harvey-gas-shortages /625106001.

20. Bradley Berman, "Electric Car Owners Unfazed by Storm," *New York Times*, November 2, 2012, http://wheels.blogs.nytimes.com/2012/11/02/electric-car-owners -unfazed-by-storm.

21. Constance Douris, "How to Charge Your Electric Vehicle (and Other Devices) When the Grid Is Down," *Forbes*, December 4, 2017, http://forbes.com/sites /constancedouris/2017/12/04/how-to-charge-your-electric-vehicle-and-other-devices -when-the-grid-is-down/#1d83c8b21559.

22. Zamuda et al., "Energy Supply," 186.

23. Ibid., 187.

24. "Framework for Long-Term Deep Carbon Reduction Planning, 2015: CNCA Cities Long Term and Interim GHG Reduction Targets," Carbon Neutral Cities Alliance, June 2014, http://usdn.org/uploads/cms/documents/cnca-framework-12-2-15.pdf.

25. "New York City's Pathway to Deep Carbon Reductions," Mayor's Office of Sustainability, City of New York, 2013, http://s-media.nyc.gov/agencies/planyc2030 /pdf/nyc_pathways.pdf; "New York City's Roadmap to 80 x 50," Mayor's Office of Sustain-ability, City of New York, 2014, https://www1.nyc.gov/assets/sustainability /downloads/pdf/publications/New%20York%20City's%20Roadmap%20to%20 80%20x%2050_20160926_FOR%20WEB.pdf; "One NYC 2017 Progress report," Mayor's Office of Sustainability, City of New York, http://onenyc.cityofnewyork.us /wp-content/uploads/2017/05/OneNYC_Progress_Report_2017.pdf; "1.5°C: Aligning New York City With the Paris Climate Agreement," Mayor's Office of Sustainability, City of New York, 2017, https://www1.nyc.gov/site/sustainability /codes/1.5-climate-action-plan.page.

26. "Climate Change and Nuclear Power 2018," International Atomic Energy Agency (2018), http://iaea.org/publications/13395/climate-change-and-nuclear-power-2018.

27. Kristin Linnerud, Torben K. Mideksa, and Gunnar S. Eskeland, "The Impact of Climate Change on Nuclear Power Supply," *Energy Journal* 32, no. 1 (January 2011): 149–68.

28. Oszvald Glöckler, "Effects of Extreme Weather on Nuclear Power Plants," Joint ICTP-IAEA Workshop on Vulnerability of Energy Systems to Climate Change and Extreme Events (April 19–23, 2010), http://bwl.univie.ac.at/fileadmin/user_upload/lehrstuhl_ind _en_uw/lehre/ws1213/SE_Energy_WS12_13/The_Impact_of_Climate_Change_on _Nuclear_Power_Supply.pdf.

29. "Technology Roadmap—Nuclear Energy 2015," International Energy Agency (2015), http://webstore.iea.org/technology-roadmap-nuclear-energy-2015.

30. Robert Watson, James McCarthy, and Liliana Hisas, "The Economic Case for Climate Action in the United States: FE-US, 2018," Universal Ecological Fund, 2, http://feu-us .org/case-for-climate-action-us.

31. Derik Broekhoff, Georgia Piggot, and Peter Erickson, *Building Thriving, Low-Carbon Cities: An Overview of Policy Options for National Governments* (Washington, DC: Coalition for Urban Transitions, 2018), http://newclimateeconomy.net/content/cities -working-papers.

U.S. CLIMATE RISK AND FINANCIAL MARKETS

1. Brad Plumer, "It's New York vs. California in a New Climate Race. Who Will Win?" *New York Times*, July 8, 2019.

2. Paul A. Griffin and Amy Myers Jaffe, "Are Fossil Fuel Firms Informing Investors Well Enough About the Risks of Climate Change?" *Journal of Energy & Natural Resources Law* 36 (2018): 381–410.

3. Russell Gold and Katherine Blunt, "PG&E's Radical Plan to Prevent Wildfires: Shut Down Power Grid," *Wall Street Journal*, April 27, 2019.

4. Coral Davenport, "Climate Change Poses Major Risks to Financial Markets, Regulator Warns," *New York Times*, June 11, 2019.

5. IPCC, "Summary for Policy Makers," in *Climate Change 2013: The Physical Science Basis*, ed. T. F. Stocker et al. (Cambridge: Cambridge University Press, 2013); T. Knutson, J. P. Kossin, C. Mears, J. Perlwitz, and M. F. Wehner, "Detection and Attribution of Climate Change," *Climate Science Special Report: Fourth National Climate Assessment*, vol. 1. (Washington: U.S. Global Change Research Program, 2017).

6. Paul Griffin, David Lont, and Martien Lubberink, "Extreme High Surface Temperature Events and Equity-Related Physical Climate Risk," Working Paper, University of California, Davis, 2019.

7. "Storm Events Database," NOAA National Centers for Environmental Information (2018), http://ncdc.noaa.gov/stormevents/ftp.jsp; "Billion-Dollar Climate and Weather Disasters," NOAA National Centers for Environmental Information (2018–2019), http://ncdc.noaa.gov/billions; "Storm Data Preparation," NOAA National Weather Service, NWSI 10-1605 (July 16, 2018): Appendix A.

8. "Final Report: Recommendations of the Task Force on Climate-Related Financial Disclosures," Task Force on Climate-Related Financial Disclosures (TFCD) (June 2017), http://fsb-tcfd.org/publications/final-recommendations-report.

9. Nicholas Silver, "Blindness to Risk: Why Institutional Investors Ignore the Risk of Stranded Assets," *Journal of Sustainable Finance and Investment* 7, no. 1 (August 2016): 99–113; Ben Caldecott, "Introduction to Special Issue: Stranded Assets and the Environment," *Journal of Sustainable Finance and Investment* 7, no. 1 (December 2016): 1–13.

10. Council on Foreign Relations, "Climate Risk Impacts on the Energy System," June 2019, https://www.cfr.org/report/climate-risk-impacts-energy-system; Ed Crooks, "Analysts Dismiss 'Carbon Bubble' Warning," *Financial Times*, October 2016, http://ft.com/content/9954e072-9587-11e6-a80e-bcd69f323a8b.

11. Other climate change–related lawsuits against fossil fuel companies focus on duty-of-care common law and the precautionary principle, in which a firm is expected to take precautionary measures when the firm's actions could threaten human health. David Hunter and James Salzman, "Negligence in the Air: The Duty of Care in Climate Change Litigation," *University of Pennsylvania Law Review* 155 (2007): 1741–94; James Cameron and Juli Aboucher, "The Precautionary Principle: A Fundamental Principle of Law and Policy for the Protection of the Global Environment," *Boston College International and Comparative Law Review* 14 (1991): 1–27; David Kriebel et al., "The

Precautionary Principle in Environmental Science," *Environmental Health Perspectives* 109 (2001): 871–76.

12.   A. Poon, "An Examination of New York's Martin Act as a Tool to Combat Climate Change" *Environmental Affairs* 44 (2017): 115.

13.   Christopher Coats, "Market Cap of U.S. Coal Companies Continues to Fall," *Institute for Energy Economics and Financial Analysis*, 2016, http://ieefa.org/market-cap-u-s-coal -companies-continues-fall.

14.   Energy Information Administration, "What Is U.S. Electricity Generation by Source?" 2019, http://eia.gov/tools/faqs/faq.php?id=427&t=3.

15.   Bankruptcies have come in several waves. Three major firms declared bankruptcy in 2015–2016. Additional bankruptcies have occurred more recently. See Kristin Lam, "Is President Donald Trump Losing His Fight to Save Coal? Third Major Company Since May Files for Bankruptcy," *USA Today*, July 3, 2019, http://usatoday.com/story/news /nation/2019/07/03/coal-collapse-third-company-may-files-bankruptcy/1644619001/; and Dana Varinsky, "Nearly Half of U.S. Coal Is Produced by Companies That Have Declared Bankruptcy—and Trump Won't Fix That," *Business Insider*, December 2019, http://businessinsider.com/us-coal-bankruptcy-trump-2016-12.

16.   Eugene Fama and Kenneth French, "Common Risk Factors in the Returns on Stocks and Bonds," *Journal of Financial Economics* 33 (1993): 3–56.

17.   High prices above \$125 per CO_2 ton would be inconsistent with empirical estimates of the implied average cost of carbon found on the stock market today as an off-the-books future financial obligation (that is, an off-balance-sheet liability) for listed companies. My research estimates the price of carbon embedded in current stock market valuations in the range of \$80 per CO_2 ton for the average U.S. firm and 75 euros for the average EU company. Paul A. Griffin, Amy Myers Jaffe, David H. Lont, and Rosa Dominguez-Faus, "Science and the Stock Market: Investors' Recognition of Unburnable Carbon," *Energy Economics* 52 (2015): 1–12; Peter M. Clarkson, Yue Li, Matthew Pinnuck, and Gordon D. Richardson, "The Valuation Relevance of Greenhouse Gas Emissions Under the European Union Carbon Emissions Trading Scheme," *European Accounting Review* 24 (2015): 551–80.

18.   Kent D. Daniel, Robert B. Litterman, and Gernot Wagner, "Applying Asset Pricing Theory to Calibrate the Price of Climate Risk," NBER working paper no. 22795, (2018), http://nber.org/papers/W22795. One estimate is that a fifteen-year delay in accurate pricing could generate economic losses of about \$10 trillion per year from the additional climate-related costs.

19.   Marco Bardoscia, Stefano Battiston, Fabio Caccioli, and Guido Caldarelli, "Pathways Towards Instability in Financial Networks," *Nature Communications* 8 (2017).

20.   Philipp H. Hildebrand and Deborah Winshet, "Adapting Portfolios to Climate Change: Implications and Strategies for All Investors," Blackrock Investment Institute (2016), http://blackrock.com/corporate/literature/whitepaper/bii-climate-change-2016-us.pdf; Stefano Battiston, Antoine Mandel, Irene Monasterolo, Franziska Schütze, and Gabriele Visentin, "A Climate Stress-Test of the Financial System," *Nature Climate Change* 7 (2017): 283.

21. J.D. Morris, "California's Strict Wildfire Liability Rule Hangs Over Bankrupt PG&E," *San Fransisco Chronicle*, February 2019, http://sfchronicle.com/business/article /California-s-strict-wildfire-liability-rule-13604239.php.

22. Russell Gold and Katherine Blunt, "PG&E's Radical Plan to Prevent Wildfires: Shut Down Power Grid," *Wall Street Journal*, April 27, 2019.

23. Sammy Roth, "PG&E's Bankruptcy Could Slow California's Fight Against Climate Change," *Los Angeles Times*, January 15, 2019, http://latimes.com/business/la-fi-pge -bankruptcy-climate-change-20190115-story.html.

24. Council on Foreign Relations, "Climate Risk Impacts on the Energy System," June 2019, http://cfr.org/report/climate-risk-impacts-energy-system.

25. Dale Kasler, "A Statewide Problem. How PG&E's Bankruptcy Could Soil California's Green-Energy Movement," *Sacramento Bee*, January 31, 2019, http://sacbee.com/news /business/article224750505.html.

26. Supran Geoffrey and Oreskes Naomi, "Assessing ExxonMobil's Climate Change Communications (1977–2014)," *Environmental Research Letters* 12 (2017); ExxonMobil Corporation v. Healey, 1:17-cv-02301 (2019).

27. Griffin and Jaffe, "Are Fossil Fuel Firms Informing Investors Well Enough About the Risks of Climate Change?" *Journal of Energy & Natural Resources Law* 36 (2018): 381–410.; Nicholas Silver, "Blindness to Risk: Why Institutional Investors Ignore the Risk of Stranded Assets," *Journal of Sustainable Finance & Investment* 7, no. 1 (2017): 99–113; Jan Bebbington, Thomas Schneider, Lorna Stevenson, and Alison Fox, "Fossil Fuel Reserves and Resources Reporting and Unburnable Carbon: Investigating Conflicting Accounts," *Critical Perspectives on Accounting* (2019).

28. Silver, "Blindness to Risk"; Caldecott, "Introduction to Special Issue."

29. ExxonMobil Corporation v. Healey, 1:17-cv-02301 (2019); Martin Finucane, "U.S. Supreme Court Refuses to Block Healey's Bid to Investigate ExxonMobil," *Boston Globe*, January 7, 2019, http://bostonglobe.com/metro/2019/01/07/supreme-court-refuses -block-healey-bid-investigate-exxonmobil/NuOpBuqIej6qzlL6X45BOM/story.html.

30. Michael G. Faure, "Insurability of Damage Caused by Climate Change: A Commentary," *University of Pennsylvania Law Review* 155 (2007): 1875–99.

31. Joanne Linnerooth-Bayer and Stefan Hochrainer-Stigler, "Financial Instruments for Disaster Risk Management and Climate Change Adaptation," *Climatic Change* 133 (2015): 85–100.

32. David Hirshleifer and Siew-Hong Teoh, "Herd Behaviour and Cascading in Capital Markets: A Review and Synthesis," *European Financial Management* 9 (2003): 25–66; Sushil Bikhchandani, David Hirshleifer, and Ivo Welch, "Learning From the Behavior of Others: Conformity, Fads, and Informational Cascades," *Journal of Economic Perspectives* 12 (1998): 151–70.

33. Paul A. Griffin, Thaddeus Neururer, and Estelle Y. Sun, "Environmental Performance and Analyst Information Processing Costs," *Journal of Corporate Finance* (2018), http://doi.org/10.1016/j.jcorpfin.2018.08.008; Aaron K. Chatterji, Rodolphe Durand, David I. Levine, and Samuel Touboul, "Do Ratings of Firms Converge? Implications for Managers, Investors and Strategy Researchers," *Strategic Management Journal* 37 (2016): 1597–614.

34. Svenn Jensen, Kristina Mohlin, Karen Pittel, and Thomas Sterner, "An Introduction to the Green Paradox: The Unintended Consequences of Climate Policies," *Review of Environmental Economics and Policy* 9 (2015): 246–65.

35. Mark Carney, "Breaking the Tragedy of the Horizon: Climate Change and Financial Stability," speech at Lloyd's of London, September 29, 2015; "Tragedy of the Horizons Briefing Note, 2 Degrees Investing Initiative," 2° Investing Initiative (2016), http://2degrees-investing.org/tragedy-of-the-horizon.

36. Nicholas Stern, "The Economics of Climate Change," *American Economic Review* 98 (2008): 1–37; Simon Dietz, Chris Hope, Nicholas Stern, and Dimitri Zenghelis, "Reflections on the Stern Review (1): A Robust Case for Strong Action to Reduce the Risks of Climate Change," *World Economics* 8 (2007): 121–68; William Nordhaus, "Critical Assumptions in the Stern Review on Climate Change," *Science* 317 (2007): 201–202; Richard S.J. Tol, "Estimates of the Damage Costs of Climate Change. Part 1: Benchmark Estimates," *Environmental and Resource Economics* 21 (2002): 47–73.

37. Phillip H. Hildebrand and Deborah Winshet, "Adapting Portfolios to Climate Change: Implications and Strategies for All Investors." Blackrock Investment Institute (2016), http://blackrock.com/corporate/literature/whitepaper/bil-climate-change-2016–is.pdf; Battison et al., "A Climate Stress-Test of the Financial System."

38. Securities and Exchange Commission, "Commission Guidance Regarding Disclosure Related to Climate Change," Release Nos. 33–9106, 34–61469, Washington, DC, February 2, 2010.

A CLEAR AND PRESENT DANGER: CLIMATE RISKS, THE ENERGY SYSTEM, AND U.S. NATIONAL SECURITY

1. Jason Samenow, "60 Inches of Rain Fell From Hurricane Harvey in Texas, Shattering U.S. Storm Record," *Washington Post*, September 22, 2017, http://washingtonpost.com /news/capital-weather-gang/wp/2017/08/29/harvey-marks-the-most-extreme-rain -event-in-u-s-history.

2. David R. Reidmiller et al., eds., *Impacts, Risks, and Adaptation in the United States: Fourth National Climate Assessment* (Washington, DC: U.S. Global Change Research Program, 2018), doi: 10.7930/NCA4.2018.CH4, 2018, http://nca2018.globalchange.gov.

3. "Hurricane Irma Cut Power to Nearly Two-Thirds of Florida's Electricity Customers," U.S. Energy Information Administration (EIA) (September 20, 2017), http://eia.gov /todayinenergy/detail.php?id=32992.

4. Thomas Gibbons-Neff, "This Is the U.S. Military's Response to Hurricane Harvey," *Washington Post*, August 28, 2017, http://washingtonpost.com/news/checkpoint/wp /2017/08/28/this-is-the-u-s-militarys-response-to-hurricane-harvey.

5. Alexia Fernández Campbell, "Puerto Rico Power Restored 11 Months After Hurricane Maria," *Vox*, August 15, 2018, http://vox.com/identities/2018/8/15/17692414/puerto -rico-power-electricity-restored-hurricane-maria.

6. Barbara Starr, Zachary Cohen, and Ryan Browne, "U.S. Military Sends Ships, Aircraft to Puerto Rico," CNN, September 26, 2017, http://cnn.com/2017/09/26/politics/us -military-response-puerto-rico-hurricane-maria/index.html.

7. Arelis R. Hernández, Whitney Leaming, and Zoeann Murphy, "Life Without Power in Puerto Rico—and No End in Sight," *Washington Post*, December 14, 2017, http://washingtonpost.com/graphics/2017/national/puerto-rico-life-without-power; Milken Institute School of Public Health, "Ascertainment of the Estimated Excess Mortality From Hurricane Maria in Puerto Rico," George Washington University and University of Puerto Rico Graduate School of Public Health (2018), http://publichealth.gwu.edu/sites/default/files/downloads/projects/PRstudy/Acertainment%20of%20the%20Estimated%20Excess%20Mortality%20from%20Hurricane%20Maria%20in%20Puerto%20Rico.pdf.

8. Megan Cerullo, "Puerto Rico Faces a Surge in Murders After Hurricane Maria," *New York Daily News*, January 11, 2018, http://nydailynews.com/news/crime/puerto-rico-faces-surge-slayings-post-maria-article-1.3752263.

9. U.S. Department of Energy (DOE), "U.S. Energy Sector Vulnerability to Climate Change and Extreme Weather," DOE/Pl-0013 (July 2013), http://energy.gov/sites/prod/files/2013/07/f2/20130716-Energy%20Sector%20Vulnerabilities%20Report.pdf.

10. Rich Heidorn Jr., "PJM Weathers Last Week's Arctic Blast," *RTO Insider*, February 3, 2019, http://rtoinsider.com/pjm-polar-vortex-cold-weather-alerts-110358; Jim Efstathiou Jr., "Deep Freeze Spawns Rare Frazil Ice That Hobbles Nuclear Reactor," Bloomberg, January 31, 2019, http://bloomberg.com/news/articles/2019-01-31/deep-freeze-spawns-rare-frazil-ice-that-hobbles-nuclear-reactor.

11. Joshua Busby, *Climate Change and National Security: An Agenda for Action* (New York: Council on Foreign Relations, November 2007), http://cfr.org/report/climate-change-and-national-security; "Who Cares About the Weather? Climate Change and U.S. National Security," *Security Studies* 17 (3): 468–504; "Climate Change and U.S. National Security: Sustaining Security Amidst Unsustainability," in *Sustainable Security: Rethinking American National Security Strategy*, ed. Jeremi Suri and Benjamin Valentino (Oxford: Oxford University Press, 2016); "Environmental Security," in *Handbook of International Security*, ed. William C. Wohlforth and Alexandra Gheciu (Oxford: Oxford University Press, 2017).

12. "Climate-Related Risk to DOD Infrastructure Initial Vulnerability Assessment Survey (SLVAS) Report," U.S. Department of Defense (DOD) (2018), http://hsdl.org/?abstract&did=807779.

13. "Report on Effects of a Changing Climate to the Department of Defense," DOD (January 2019), http://climateandsecurity.files.wordpress.com/2019/01/sec_335_ndaa-report_effects_of_a_changing_climate_to_dod.pdf; the U.S. Government Accountability Office (GAO) also reviewed the climate resilience of overseas DOD installations ("DOD Needs to Better Incorporate Adaptation Into Planning and Collaboration at Overseas Installations," GAO-18-206 (November 2017), http://gao.gov/products/GAO-18-206).

14. Alejandro Quiroz Flores and Alastair Smith, "Surviving Disasters," working paper, New York University, 2010.

15. Maxwell Strachan, "The Definitive History of 'George Bush Doesn't Care About Black People,'" *Huffington Post*, August 28, 2015, http://huffingtonpost.com/entry/kanye-west-george-bush-black-people_us_55d67c12e4b020c386de2f5e.

16. Angela Fritz, "Hurricanes Harvey, Irma, Maria and Nate Were So Destructive, Their Names Have Been Retired," *Washington Post*, April 12, 2018, http://washingtonpost.com/news/capital-weather-gang/wp/2018/04/12/hurricanes-harvey-irma-maria-and-nate-were-so-destructive-their-names-have-been-retired.

17. Shan Carter and Amanda Cox, "One 9/11 Tally: $3.3 Trillion," *New York Times*, September 8, 2011, http://archive.nytimes.com/www.nytimes.com/interactive/2011/09/08/us/sept-11-reckoning/cost-graphic.html.

18. Helene Cooper, Michael D. Shear, and Denise Grady, "U.S. to Commit up to 3,000 Troops to Fight Ebola in Africa," *New York Times*, September 15, 2014, http://nytimes.com/2014/09/16/world/africa/obama-to-announce-expanded-effort-against-ebola.html.

19. Barack Obama, "Remarks by the President on the Ebola Outbreak," September 16, 2014, http://obamawhitehouse.archives.gov/the-press-office/2014/09/16/remarks-president-ebola-outbreak.

20. Thomas Beller, "The Residents of This Louisiana Island Are America's First 'Climate Refugees,'" *Smithsonian*, June 29, 2016, http://smithsonianmag.com/science-nature/residents-louisiana-island-americas-first-climate-refugees-180959585.

21. Vaughn Hillyard, "Evacuated Residents of Irma-Wracked Barbuda Consider Not Going Back," NBC News, September 17, 2017, http://nbcnews.com/storyline/hurricane-irma/evacuated-residents-irma-wracked-island-barbuda-consider-not-going-back-n802141.

22. DOE, *Quadrennial Energy Review: Transforming the Nation's Electricity System: The Second Installment*, Washington (2017), http://energy.gov/epsa/quadrennial-energy-review-second-installment.

23. Rose Thayer and Corey Dickstein, "How Florence Is Affecting Area Bases," *Stars and Stripes*, September 15, 2018, http://stripes.com/how-florence-is-affecting-area-bases-1.547531.

24. "Military Expert Panel Report: Sea Level Rise and the U.S. Military's Mission," Center for Climate and Security (September 2016), http://climateandsecurity.org/militaryexpertpanel.

25. Christopher Flavelle, "FEMA Strips Mention of 'Climate Change' From Its Strategic Plan," Bloomberg, March 15, 2018, http://bloomberg.com/news/articles/2018-03-15/fema-strips-mention-of-climate-change-from-its-strategic-plan; "2018–2022 Strategic Plan," Federal Emergency Management Agency (2018), http://fema.gov/media-library-data/1533052524696-b5137201a4614ade5e0129ef01cbf661/strat_plan.pdf.

26. Reidmiller et al., eds., *Impacts, Risks, and Adaptation*.

27. Thomas Homer-Dixon, "The Rise of Complex Terrorism," *Foreign Policy*, November 16, 2009, http://foreignpolicy.com/2009/11/16/the-rise-of-complex-terrorism.

28. "Climate Change and the Electricity Sector: Guide for Assessing Vulnerabilities and Developing Resilience Solutions to Sea Level Rise," U.S. Department of Energy (DOE) (July 2016), http://energy.gov/sites/prod/files/2016/07/f33/Climate%20Change%20and%20the%20Electricity%20Sector%20Guide%20for%20Assessing%20Vulnerabilities%20and%20Developing%20Resilience%20Solutions%20to%20Sea%20Level%20Rise%20July%202016.pdf.

29. DOE, *Quadrennial Energy Review*.

30.   Samantha Harris and David Wei, "Why Climate Resilience and Supply Chains Go Hand in Hand," BSR, September 10, 2018, http://bsr.org/en/our-insights/blog-view /climate-change-supply-chains-go-hand-in-hand.

31.   "Report on Effects of a Changing Climate to the Department of Defense," U.S. Department of Defense (DOD) (January 2019), http://climateandsecurity.files. wordpress.com/2019/01/sec_335_ndaa-report_effects_of_a_changing_climate_to _dod.pdf; "DOD Needs to Better Incorporate Adaptation Into Planning and Collaboration at Overseas Installations," GAO-18-206 (November 2017), http://gao .gov/products/GAO-18-206).

32.   The U.S. Department of Energy has done some analysis of sea-level rise for nine metropolitan areas, including Norfolk, Los Angeles, Houston, Boston, Baltimore, Mobile, New York, Philadelphia, and Miami (DOE, "Climate Change and the Electricity Sector"; see also "Sea Level Rise and Storm Surge Effects on Energy Assets for Select Major Metropolitan Areas," DOE, http://arcgis.com/apps/MapSeries/index.html?appid =58f90c5a5b5f4f94aaff93211c45e4ec). The EIA has mapped energy infrastructure susceptible to coastal and inland flooding ("EIA Mapping Tool Shows Which U.S. Energy Facilities Are in Areas at Risk of Flooding," *Today in Energy*, August 6, 2014, http://eia.gov/todayinenergy/detail.php?id=17431). DOE has also carried out regional analysis of the main vulnerabilities with private-sector partners ("Climate Change and the U.S. Energy Sector: Regional Vulnerabilities and Resilience Solutions," U.S. DOE (2015), http://energy.gov/sites/prod/files/2015/10/f27/Regional_Climate_Vulnerabilities _and_Resilience_Solutions_0.pdf). DOE has also contracted a number of academic studies to assess multi-hazard risks to the electricity sector (Benjamin L. Preston, Scott N. Backhaus, Mary Ewers, et al., "Resilience of the U.S. Electricity Sector: A Multi-Hazard Perspective," U.S. Department of Energy (2016), http://energy.gov/sites/prod /files/2017/01/f34/Resilience%20of%20the%20U.S.%20Electricity%20System %20A%20Multi-Hazard%20Perspective.pdf).

33.   "Climate Change and Rapidly Intensifying Hurricanes," Climate Central (May 30, 2018), http://climatecentral.org/gallery/graphics/climate-change-and-rapidly -intensifying-hurricanes; Peter Sousounis, "Why Climate Change and Hurricane Stalls Mean Flooding Rain," Verisk (September 13, 2018), http://air-worldwide.com/Blog /Why-Climate-Change-and-Hurricane-Stalls-Mean-Flooding-Rain.

ACKNOWLEDGMENTS

Richard N. Haass, president of the Council on Foreign Relations, refers to climate change as the "quintessential global challenge" and potentially the defining issue of this century. Not only can no single country solve climate change on its own, but, as Haass notes, there is no way for any single country to shield itself from its effects. The United States is no exception. When I joined the Council on Foreign Relations in 2017, Haass and CFR Senior Vice President and Director of Studies James M. Lindsay offered their strong support to me to expand CFR's work in the area of energy and climate change. I extend my deepest thanks for this opportunity and for their forward-looking vision and specific support for this project to study the financial, security, and technological dimensions of climate risk to the U.S. energy system. The subject needs more careful attention in the national debate.

I am deeply indebted to the Alfred P. Sloan Foundation, which has supported my research for many years. The workshop that served as the focal point for the research and policy analysis contained in this collection was the seventh in a series of energy-focused meetings supported by the Sloan Foundation at CFR. I am incredibly grateful for the contribution of Evan S. Michelson who, through his work at the Sloan Foundation and his valuable insights into the links between energy markets and climate change, helped shape the focus of this meeting and the resulting discussions. A vast body of knowledge and academic expertise has been created over the years through Evan's tireless efforts and the United States' energy policy is the richer for it.

I would like to thank the volume's contributors who presented at the workshop and subsequently worked with me to put together this manuscript. These authors are more than just top specialists in their respective academic fields; they are civic scientists who actively contribute their time and expertise for the betterment of society. They do not author articles in a vacuum but rather serve as an ongoing resource to policymakers in regions of the United States where the effects of climate change are bringing real and dramatic outcomes to the daily lives of many Americans. I know this firsthand because, in earlier years, I lived through severe hurricanes, flooding, and power outages with my wonderful neighbors in Houston and have been stranded on my daily commute in recent years by fires in Northern California. This is not just a theoretical exercise for me, just as it is not for millions of other Americans.

This essay collection would not have been possible without the many insights of the diverse collections of experts, government leaders, and businesspeople during the March 2019 meeting, where the free exchange of ideas hopefully will help shape the discussion of climate risk in the future. I would especially like to extend my deepest thanks to Council on Foreign Relations Research Associate Benjamin Silliman for his dogged determination to keep this programming humming smoothly regardless of the unusual circumstances that can arise in today's world with each new day. He made valuable contributions to the content of this work. Finally, my heartfelt appreciation to the CFR editorial team of Patricia Dorff, Chloe Moffett, Sumit Poudyal, and Julie Hersh, who work tirelessly to ensure that our knowledge is as readable and accessible as possible, and to CFR's events, digital, media, membership, and finance teams that work diligently behind the scenes to ensure that our efforts are successful.

Amy Myers Jaffe
August 2019

ABOUT THE AUTHORS

Jim Blackburn is an environmental lawyer and professor in the practice of environmental law in Rice University's Civil and Environmental Engineering Department. At Rice, Blackburn is a faculty scholar at the Baker Institute and codirector of the Severe Storm Prevention, Education, and Evacuation from Disasters (SSPEED) Center, and directs the undergraduate minor in energy and water sustainability. He is the author of two books, *Book of Texas Bays* and *A Texan Plan for the Texas Coast*. He was designated a Rice University distinguished alumni laureate in 2018, won the International Crane Foundation Good Egg Award in 2015, the Barbara C. Jordan Community Advocate Award from Texas Southern University in 2007, and the Robert C. Eckhardt Lifetime Coastal Achievement Award from the General Land Office of Texas in 1998, among other awards. He received a BA in history and a JD from the University of Texas at Austin and an MS in environmental science from Rice University.

Joshua Busby is an associate professor of public affairs at the University of Texas at Austin's Lyndon B. Johnson School of Public Affairs. He is also a senior research fellow with the Center for Climate and Security. The author of two books on social movements, he has written extensively on climate change and national security in publications such as *Foreign Affairs, International Security, Political Geography, Security Studies*, and *World Development*, as well as for several think tanks. One of the lead researchers on a project funded by the Department of Defense, Climate Change and African Political Stability, he was also the principal investigator of another Department of Defense–funded project, Complex Emergencies and Political Stability in Asia. Busby was a research fellow at Princeton University's Woodrow Wilson School of Public and

International Affairs, the Harvard Kennedy School, and the Brookings Institution. He earned his PhD from Georgetown University.

Christina Copeland leads CDP Worldwide's water work in North America, engaging with companies to improve their disclosure, internal governance, and the robustness of their water targets and actions in the communities where they operate. A frequent public speaker, she also works on the topic of collective action and the importance of convening government entities, local communities, companies, investors, and civil society organizations to address and mitigate shared water challenges at a local level. Prior to joining the CDP water security team, Copeland led the organization's corporate climate change and water disclosure work for materials companies and executed CDP's water membership program in the United States and Canada. Named to the GreenBiz 2016 "30 Under 30" list, she serves on the board of advisors for Yale University's Dwight Hall Socially Responsible Investment Fund. Previously, Copeland worked in Mumbai, India, doing corporate sustainability consulting for Tata Consultancy Services. She received a BS in environmental science from Cornell University.

Paul A. Griffin is a distinguished professor at the Graduate School of Management at the University of California, Davis, and a leading international authority in accounting, financial information, and corporate disclosure. He has published more than seventy articles in leading accounting and finance journals, five research monographs for the Financial Accounting Standards Board, and two case books on U.S. corporate financial reporting. His more recent work studies how financial markets process nonfinancial information, specifically, how asset prices respond to greenhouse gas emissions, extreme high–surface temperature events, reports of stranded assets on company balance sheets, and climate risk disclosures. Articles in 2018 addressed the effect of corporate social responsibility (CSR) performance ratings on analysts' information processing costs, the effects of social norms on CSR disclosure, the sufficiency of fossil fuel firm risk disclosures, and how a firm's innovative effectiveness can improve its credit rating. Griffin recently served as coeditor of *Accounting Horizons*, a leading academic journal of the American Accounting Association. He earned an MA in operations research and economic theory and a PhD in accounting from Ohio State University.

Amy Myers Jaffe is the David M. Rubenstein senior fellow for energy and the environment and director of the Energy Security and Climate

Change program at the Council on Foreign Relations. A leading expert on global energy policy, geopolitical risk, and energy and sustainability, Jaffe previously served as executive director for energy and sustainability at the University of California, Davis, and senior advisor for energy and sustainability at the office of the chief investment officer of the University of California, Regents. She was also formerly a global fellow at the Woodrow Wilson International Center for Scholars. Prior to joining the University of California, Davis, Jaffe was founding director of the Energy Forum at Rice University's James A. Baker III Institute for Public Policy and its Wallace S. Wilson fellow for energy studies. She has taught energy policy, business, and sustainability courses at Rice University, University of California, Davis, and Yale University.

Sara Law is vice president for global initiatives at CDP Worldwide, where she works in close partnership with many entities, including businesses and investors, strategic partners such as the World Bank Group and various UN agencies, and a coalition of progressive climate change organizations under the banner We Mean Business. Her portfolio also includes carbon pricing, which is increasingly recognized as a helpful tool in the transition to a low-carbon economy. Before joining CDP in 2013, Law worked for several years in Australia's private and public legal spheres, specializing in complex commercial litigation and government relations. She is a member of the bar of the State of New York and the Supreme Court of Victoria, Australia. Law earned a BA in history and international politics and an LLB from the University of Queensland in Australia.

Joan M. Ogden is professor emeritus of environmental science and founding director of the sustainable transportation energy pathways program at the University of California, Davis. Prior to joining the University of California, Davis, in 2003, she was a research scientist at Princeton University's Environmental Institute. Her primary research interest is techno-economic assessment of low-carbon energy technologies, especially alternative fuels, hydrogen, fuel cells, and renewable energy. Her recent work centers on hydrogen infrastructure strategies and applications of fuel cell technology in transportation and stationary power. Ogden has served on high-level committees advising clean energy policies across the United States and internationally. The author of two books and numerous technical articles, she holds a BS in mathematics from the University of Illinois and a PhD in theoretical physics from the University of Maryland.

Impact of Climate Risk on the Energy System

www.ingramcontent.com/pod-product-compliance
Lightning Source LLC
Chambersburg PA
CBHW060516280326
41933CB00014B/2990